ENCYCLOPEDIA OF ANIMALS

Published in Great Britain by Cliveden Press,
An Egmont Company, Egmont House, P.O. Box 111,
Great Ducie Street, Manchester M60 3BL.
Printed in Singapore. ISBN 7235 1681 2

CLIVEDEN PRESS

Contents

Above: Freshwater insects are eaten by small fishes such as minnows. They in turn are eaten by larger fishes such as trout, as well as by birds like the kingfisher. This is a simple food chain.

The Balance of Nature

In the wild, plants and animals depend on each other for survival. Most plants make their own food with the help of energy from the sun. Animals, on the other hand, must eat plants or other animals to obtain the energy they need. The process by which energy is passed from one creature to another is called a food chain. Thus plants use the sun's energy to produce food from simple materials. Animals called herbivores eat the plants, and they themselves are eaten by carnivorous animals. If the numbers in any part of the chain rise or fall dramatically, this affects other links in the chain. Thus if a crop fails, the herbivores starve and fewer predators will survive. This is part of the process called the balance of nature.

Food chains are not always as simple as this. The carnivore may eat many types of prey and may itself be eaten by several other types of predator. When several food chains become interconnected in this way they form a food web.

CARBON CYCLE

All living things contain carbon. As much as 18 per cent of all living material is carbon, so it is a very important chemical. Carbon is never lost but is always recycled.

All living things breathe out carbon dioxide, and plants use the gas to make food when they photosynthesize. When the plant is eaten by an animal, the carbon is taken into its body. Later on, the animal returns some of the carbon to the atmosphere when it breathes out carbon dioxide. When an animal or plant dies, the carbon in its body is again returned to the atmosphere, this time through the action of bacteria. In this way the carbon is recycled through the environment and very little is lost.

Below: When rabbits eat plants they take in carbon but, like all living things, they release it as carbon dioxide when they breathe out. Plants take carbon dioxide in again when they photosynthesize, and in this way the carbon is recycled.

Animal Populations

The numbers of animals and plants remain fairly constant from year to year. There are several factors which keep their populations in check. First the amount of food available is often limited, so the more animals there are, the less food there is to go round. If the numbers of an animal suddenly increase for some reason then the numbers of its predators soon increase as well. This gradually brings the numbers down again.

Sometimes the controls fail and numbers of a species increase unchecked. If they affect man they are called pests. In many cases man has upset the balance of nature and brought about the population explosion without realizing it.

Many insects feed on plants, but will often feed on only a few species. In the wild, the plants are generally spaced out. This spacing out helps to keep the insect population at a steady level.

But if the insect finds an area where its foodplant is abundant its numbers will soon rise. Man often grows vast areas of crops for food. As a result those insects that can eat the crop benefit and soon become pests.

Above: The Colorado beetle can become a serious pest of man's crops. It is particularly fond of potatoes and if not controlled can devastate a whole harvest.

Man often controls pests by spraying them with chemicals. But sometimes he can use natural agents as a form of biological control. Rabbits have been controlled by the introduction of a disease called myxomatosis which is transmitted by fleas. In recent years, however, many rabbits have become immune to myxomatosis.

Ichneumon wasps also make good agents of control since they lay their eggs in other insects and eventually kill them.

Right: The mongoose is skilled at catching wary prey such as the black rat. It helps keep down the numbers of rats and acts as a natural 'control' agent.

Above: Charles Darwin was born in 1809, the son of a Shrewsbury doctor. Having abandoned careers in medicine and the church, he joined HMS *Beagle* as ship's naturalist. His voyages took him all round the world and helped inspire his theories of evolution.

Evolution in Action

Above: During his visit to the Galapagos, Darwin observed remarkable variation in the finches. On the widely separated islands there were separate races. Darwin suggested that all the races had evolved from the same original stock, a species from South America, and that each had adapted to the particular conditions on each island.

Nowadays, evolution is an accepted fact, but it has not always been so. Once, people thought that life arose spontaneously from things like mud, water and air. The Biblical idea of creation was also widely accepted until scientists like Charles Darwin began to introduce new ideas.

Darwin's studies led him to believe that the present-day variety of life on earth had developed from much simpler creatures which lived long ago but had gradually died out, being replaced by more complex ones which were better suited to their environment. Only the most successful animals and plants survived and passed their characters on to their offspring. His ideas were criticized and ridiculed at the time, especially by the church, but he is now known to have been correct about many of his ideas concerning evolution.

Below: Although no-one would confuse the adult forms of man, chicken and dogfish, it would take an expert to identify their embryos. Darwin suggested that the reason for their similarities was that they are all descended from a common ancestor and, during their development, they go through all the stages of evolution. All vertebrate embryos share many similar features as they develop, looking very much alike at some stages. At an early stage all embryos have gill pouches on their heads, a feature only retained in the adult form by fishes.

Studying Evolution

Of the countless millions of animals and plants which have lived and died on this earth before us, a small number died in places where their remains could be preserved as fossils until the present day. Some animals and plants lend themselves to fossilization because of their tough bodies. This is why there is an abundance of fossil pollen, and numerous species of shellfish. By examining changes in the fossil species and relating these to the geological time scale, we can begin to piece together the story of evolution – a process which still goes on today.

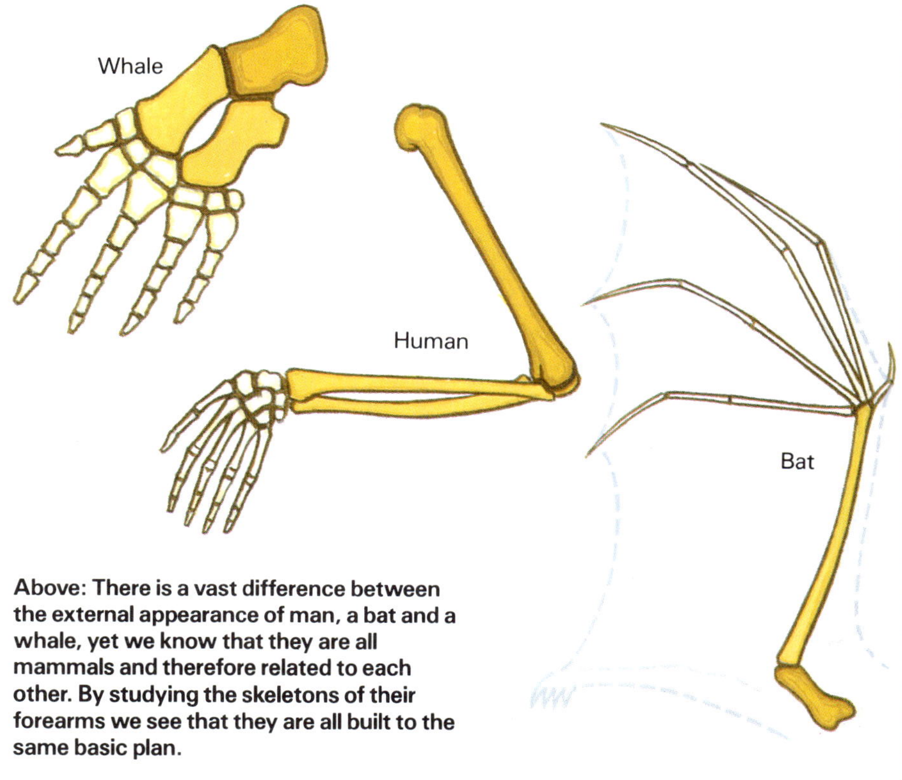

Above: There is a vast difference between the external appearance of man, a bat and a whale, yet we know that they are all mammals and therefore related to each other. By studying the skeletons of their forearms we see that they are all built to the same basic plan.

After travelling widely throughout the world and making detailed observations abroad and at home on domesticated animals, Darwin proposed that all life on earth had evolved gradually from the simplest forms over a vast period of geological time. As evidence for his theories he used fossil records, the 'artificial selection' of breeders of domestic pigeons, and his observations on adaptations of species on isolated islands.

In his book 'The Origin of Species', published in 1859, he explained his ideas about natural selection. Animals and plants produce a vast number of offspring in each generation, far more than are needed to maintain a stable population of the species. Most of the young die before they reach adulthood. Darwin proposed that only those most able to survive would do so, the rest would be 'selected out' by nature.

The Animal Kingdom

The Animal Kingdom can be divided simply into two large groups: animals without backbones (also known as invertebrates) and animals with backbones (also known as vertebrates).

Some of the more familiar invertebrate animals we see around us include insects, crabs, worms and spiders. Animals like cats and dogs, snakes, frogs, birds and fishes belong to the other group, the vertebrates.

Vertebrate animals are generally much bigger than invertebrate animals. They also *seem* to be more common. However, there are many more invertebrate animals in the world than there are vertebrate animals. About 95 per cent of the whole Animal Kingdom is composed of invertebrate animals. Many of them spend much of their lives hidden from view, however — in the soil, in the sea or concealed among the bark of trees.

Animal Classification

Animals (and plants) are grouped together according to particular features that they have in common. We have already seen how animals can be grouped into vertebrates and invertebrates. The next major division within the Animal Kingdom is the phylum. Animals within a phylum also share many features in common. For instance, the animals grouped together in the Phylum Arthropoda all have a tough, external skeleton called an exoskeleton, and jointed legs.

Within each phylum the animals are arranged in separate classes. Animals within each class share many common features of body structure. Thus, within the Phylum Arthropoda, the Class Insecta includes all those animals whose bodies are divided into three parts and bear three pairs of legs — in other words, the insects.

The next division, within the classes, is the order. Within each order are usually a number of families. Families contain animals which are similar in many ways. In the world of birds, all the tits (blue tit, coal tit and so on) are grouped together in the same family.

The next division is the genus. The final division is the species. All the species within a particular genus are basically alike. It is only animals of the same species breeding together which can produce young animals capable of themselves breeding.

Right: This family tree of the Animal Kingdom shows the main groups of animals which have appeared on earth.

Left: Although they all look quite different, all these animals are mammals. Their bodies bear fur and they suckle their young.

Crustaceans
Segmented worms
Flatworms
Echinoderms
Fishes
Birds
Molluscs
Roundworms
Insects
Reptiles
Mammals
Amphibians
Arachnids
Protochordates
Bryozoans
Sponges
Protozoa
Comb jellies
Jellyfishes

World Wildlife

About 200 million years ago, when dinosaurs roamed the Earth, the continents were joined together in a vast supercontinent called Pangaea. But slowly the continents drifted apart and animals living on one land mass became separated from those living on another. Each group of animals continued to change – to *evolve* – but, isolated from each other, they evolved in different ways. This is why, today, certain animals are found only on certain continents; why, for example, kangaroos are found only in Australia and llamas only in South America. The map shows animals typical of each area.

NORTH AMERICA

Only a narrow stretch of water separates North America from Asia. Often in the past the two land masses have been joined by a land bridge, and animals have been able to pass from one to the other. For this reason, many of the animals found in North America are closely related to those found in Asia and Europe.

SOUTH AMERICA

South America has many animals found nowhere else in the world. Some are quite strange in both appearance and behaviour, such as the sloth, spending its days hanging upside-down in the trees, or the toothless anteater, rummaging through ant hills with its long nose. There is also a wealth of beautiful birds such as the scarlet macaw in the tropical rain forests.

EUROPE AND ASIA

Many people live in Europe and Asia, and much of the land is farmed, so the larger wild animals are mostly confined to nature parks. But some animals, such as the fox, have adapted to living among people. Others, such as the giant panda, live only in the deepest parts of the forest.

AFRICA

The grasslands of Africa have the most varied collection of large grazing animals in the world – including the giraffe, elephant, rhinoceros, antelopes and many others. These animals and the carnivores which prey on them roam free in vast national parks.

AUSTRALASIA

Australia was the first continent to break away from Pangaea. It did so at a time when most of the world's mammals reared their young in pouches. These animals, called marsupials, flourished in Australia, but soon became extinct in other parts of the world. The koala and kangaroo are both marsupials, and are found only in Australia. The tuatara, a reptile found in New Zealand, is the nearest living relative of the great dinosaurs.

Animals of Long Ago

The first animals appeared on earth many millions of years ago. Their remains sometimes become preserved in rocks. When this happens the remains are known as fossils. By studying fossils, scientists have discovered almost everything they know about the animal life of long ago.

Life began in the oceans. The first forms of life were small, primitive creatures rather like the sponges and jellyfishes of today. In time, larger creatures such as crabs evolved. Then fishes appeared in the seas. Some of these fishes were huge, armour-plated monsters which preyed on other sea creatures.

Eventually fishes evolved which could breathe air. From them developed animals which were able to live both on land or in the water. These were the amphibians, the ancestors of the salamanders, frogs and toads. They first appeared about 350 million years ago.

Once the land had been conquered by animals, other creatures evolved. Next to arrive were the reptiles. For many millions of years, conditions on earth were exactly right for the reptiles, and they flourished in many parts of the world.

Right: *Pteranodon* was a huge flying reptile that lived during the Cretaceous Period, some 130 million years ago. It had a wingspan of about 7 metres. *Pteranodon* probably clambered up on to cliffs and launched itself into the air to soar and glide over the sea. *Pteranodon* had a long bony crest at the back of its head to help counter-balance the huge beak.

Far right: *Alamosaurus* lived in parts of Europe and North America during the Cretaceous Period. It was a large plant-eating dinosaur. *Tyrannosaurus* (the name means 'tyrant lizard') is shown here attacking *Alamosaurus. Tyrannosaurus* was the biggest carnivorous reptile ever to have lived. It stood over 6 metres tall.

Left: About one and a half million years ago, great ice sheets from the North Pole crept over many parts of the northern hemisphere. To keep warm, animals such as the woolly mammoth shown here grew long coats. Other shaggy coated animals included the woolly rhinoceros and the cave bear. The sabre-toothed tiger was one of several meat-eating animals which preyed on creatures at this time.

Pteranodon Alamosaurus Tyrannosaurus

The biggest reptiles were the dinosaurs. Dinosaurs ruled the earth for about 130 million years. Giant plant-eating reptiles wallowed in the swamplands, browsing on vegetation. On dry land herds of horned dinosaurs roamed about. Many of these plant-eating dinosaurs fell prey to ferocious carnivores like *Tyrannosaurus*.

Some reptiles took to the air, gliding about on leathery wings. Others returned to the sea to feed on fishes and other marine creatures. About 65 million years ago, however, most of the reptiles suddenly died out.

The Age of Mammals

By the time most of the reptiles had died out, the mammals and birds had already evolved. The mammals quickly filled all the habitats occupied by the reptiles. The first mammals were small, mouse-like animals. In time many other kinds of mammal also appeared. Some of these were plant-eaters and others became carnivorous, feeding on the plant-eaters. Today, the mammals are the dominant animals on earth.

Scientists have found fossil bones from many of these early mammals. We know that some were gigantic. There were ground sloths bigger than elephants, and a giant rhinoceros called *Baluchitherium* that was nearly as tall as a house.

Below: Some animals have become extinct in more recent times. The dodo, a bird about the size of a turkey, was once found on Mauritius island in the Indian Ocean. When sailors from visiting ships landed on the island they stole its eggs and killed the birds for food. The last dodo was killed little more than 200 years ago.

Nature Trails

The best way to find out about animals and plants is to take a walk in the country, or even take a look in a town garden.

Many national parks and country parks have nature trails marked out to lead naturalists to the places where they can see the most interesting plants and animals. They often provide identification plates similar to those opposite, to help people to put a name to what they see.

Most naturalists following nature trails find that it is a good plan to take a notebook and pencil, to make a record of anything they see, and even a sketch to help identification later. Binoculars are a great help for studying birds, and a pocket magnifying glass is essential to find out the details of plants which may look alike to the casual eye.

There is another form of nature trail, too – that left by the feet of animals as they go about their daily lives. Such footprints show up best in snow, but they can also be detected in soft ground, and even in dewy grass. There are other marks, too, that betray the presence of animals, such as the claw marks which badgers leave on the trees and logs they use as scratching posts, or holes in a bank which are home to some small creatures.

The best time to see and identify tracks left by animals is when snow is on the ground. Remember animals leave slightly different marks according to the way they are moving, and this shows up clearly in snow.

Oak
Lombardy poplar
Birch
Scots pine
Larch
Poppy
Meadow thistle
Lesser celandine
Common violet
Cuckoo-pint
Tawny owl
Swallow
Redstart
Lapwing
Pochard
Mouse, bounding
Rabbit, hopping slowly
Stoat, bounding
Squirrel, jumping

Nature in the Garden

It is not always necessary to go on a nature trip to the countryside, or to the seashore, in order to find plenty of wildlife. For many of our own back gardens provide a wonderful haven for plants and animals.

Some gardens are better than others for attracting wildlife. The best ones are those with plenty of trees growing in them, and where a part of the garden is allowed to grow 'wild'. This can easily be done by sowing some wild flower seeds in a small corner, or allowing a little of the lawn to grow longer than usual.

A pond will also prove to be a welcome feature for wild animals, many of which will come to drink, feed or lay their eggs. A small pile of rocks of even just a few bricks will make a home for spiders, slow worms and other small creatures.

However, even if your garden is a neat, tidy place without any trees, there will always be some wildlife present. All gardens

have soil, of course, and this is where many of the smaller creatures of the garden live. Earthworms are the gardener's friend, for they help mix the earth together, and create air passages for the roots of plants. Other animals of the soil include beetles and their larvae, centipedes and millipedes, and ants. If you carefully lift a large stone in the garden you will often see ants scuttling about, for they often build their nests under stones.

Sometimes gardeners find large heaps of soil on their carefully tended lawns. These are caused by moles, which burrow through the earth and push the soil they have excavated up on to the grass.

Hidden among the fallen leaves or creeping between the stems of the flowers in the flower beds are other animals such as slugs and snails.

Many of the creatures of the garden are very secretive. They remain hidden from view and only emerge from the safety of their hiding places to feed at night. Birds, however, are among the easiest of garden animals to see. We can encourage them by planting trees, or by growing bushes with berries which they can eat in the autumn. In winter, a bird feeding table will also attract hungry flocks, eager for something to eat.

Wild plants, too, will find their way into the garden. The plants we call weeds are really just wild flowers growing uninvited. In autumn, mushrooms and toadstools will appear mysteriously from the ground, and lichens, mosses and other lowly plants will soon colonize rocks.

Above: You can easily make a wormery yourself. The one shown here is made from two sheets of glass with soil sandwiched between them, but a jam jar containing some moist soil and a few leaves is just as good. Watch the earthworms mix the soil as they burrow through it.

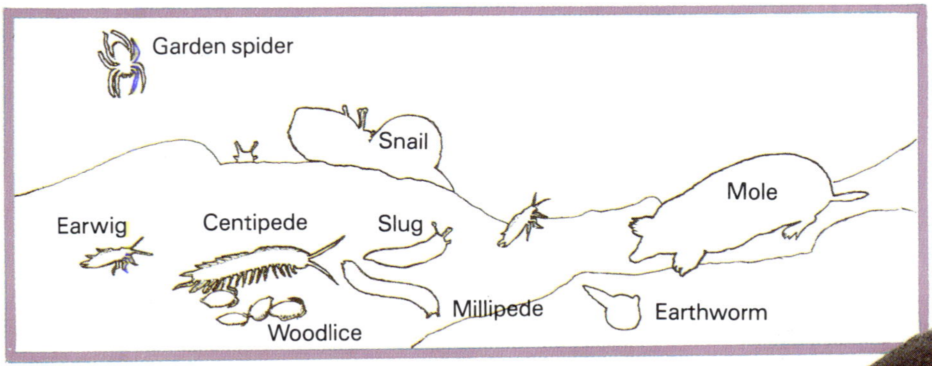

Below: Some common garden animals. Many of the smaller creatures spend the day hidden under stones or leaves, but at night they are much more active and can easily be seen with the aid of a torch.

The World of Insects

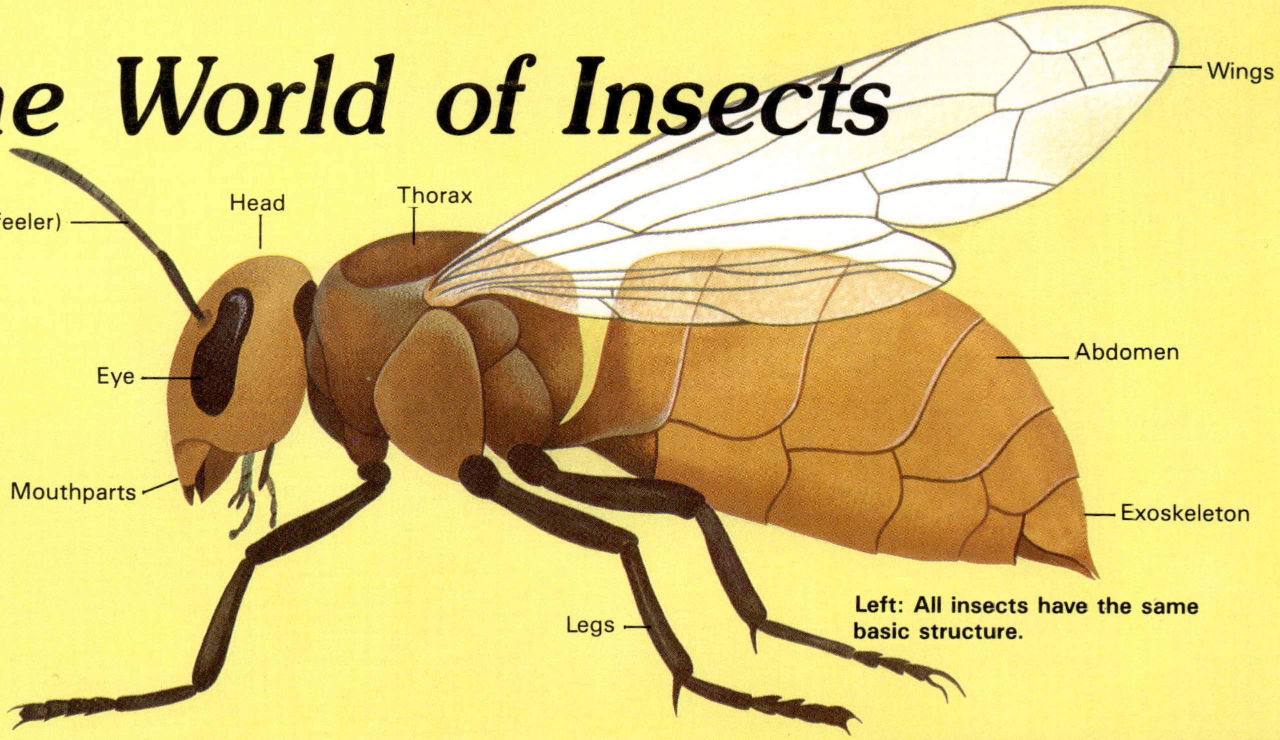

Left: All insects have the same basic structure.

Insects are the most numerous and widespread group of animals on earth. There are more species of insect than all the rest of the species in the Animal Kingdom put together. Over a million different kinds have been discovered. Insects are found in almost every sort of habitat, from deserts to lakes, and from the soil to the air. The only place which insects have been unable to colonize successfully is the oceans, but they make up for this by their abundance elsewhere. Insects vary enormously in their general shape as well as in their size. They range from the microscopic to beetles the size of a large mouse.

Despite the enormous variety of insect life, they all have certain features in common. Like many of their invertebrate relatives, insects have an external skeleton called an exoskeleton which provides support and protection. The bodies of adult insects are divided into three parts: the head, the thorax and the abdomen, and on the middle part are found three pairs of jointed legs.

Insect Lives

Perhaps the most striking feature of insects is their ability to fly. It is the conquest of the air which has helped insects to become so successful. Some insects, such as beetles, are relatively cumbersome in flight and can cover only short distances. Others, however, such as dragonflies, spend virtually all the hours of daylight on the wing. Their mastery of the air, and their excellent vision, allow them to catch even the speediest of airborne prey.

The wings of many butterflies are extremely colourful and are used for a variety of purposes such as display, defence and camouflage, in addition to their use in flight.

Insects eat many different types of food. Many, like the antlion larva above, are carnivores while others eat only plant material. Some insects even eat plants at one stage in their life and animals in another. Others are scavengers, or attack man's clothes, books and carpets. Even more extraordinarily some insects, such as certain moths, do not feed at all as adults.

Above: The antlion is a curious insect which is found in dry, sandy soils. The larva is a fierce carnivore with a liking for ants, but it is so slow and cumbersome that it has to use cunning means to catch its food. Instead of chasing its prey, the antlion larva buries itself in the sand and digs a pit. Any unfortunate ant which falls over the edge of the pit slowly but surely slides down into the jaws of the waiting larva.

Insects have well-developed senses that tell them everything they need to know about their surroundings. Unlike our own eyes, the 'compound' eyes of insects are composed of up to 25,000 separate chambers each with a small lens. Although giving

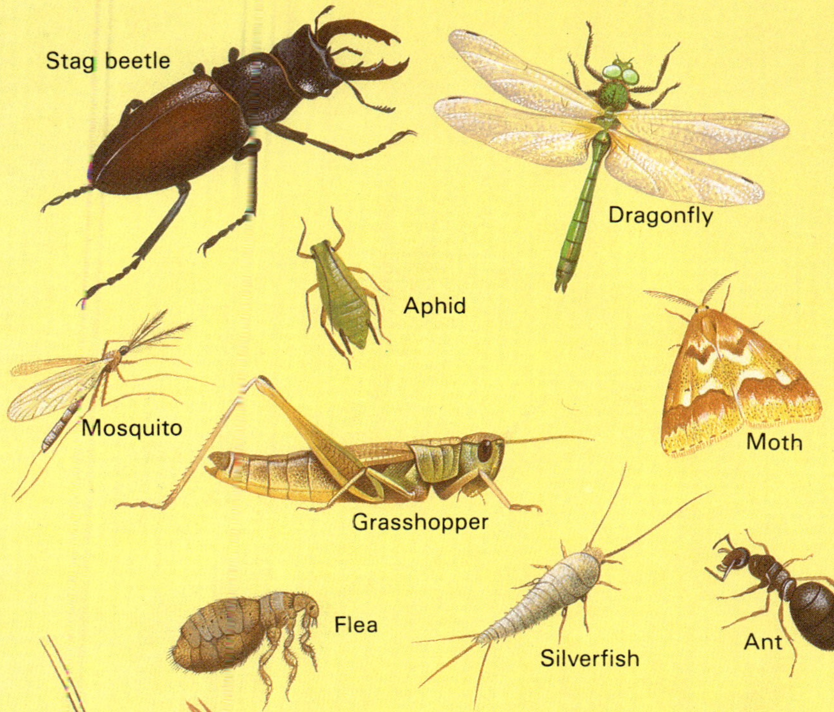

Here are some of the many different kinds of insects in the world. The vast array of insects can be subdivided into 29 main groups known as orders. Although all the members of these orders share the main features of insects, such as six true legs and a three-part body, in some species other features may be missing. For example, the parasitic insects such as fleas and lice have lost their wings. In the beetles the first pair of wings form a wingcase inside which the true wings are folded.

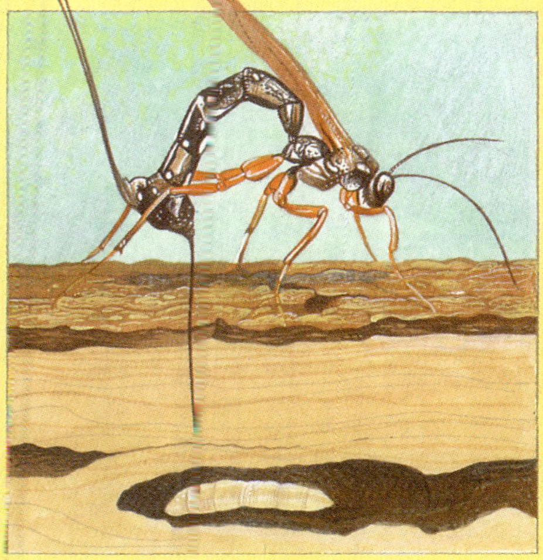

Above: Ichneumon wasps are fascinating insects which specialize in laying their eggs inside the bodies of other insect larvae. Since most insect larvae have soft skins this does not present many problems. But when the larva lives inside wood, difficulties can arise! Amazingly, some female ichneumon wasps can detect larvae buried up to several centimetres inside a tree. They then force a thin egg-laying tube, like a needle, through the wood to lay their eggs. When the ichneumon wasp eggs hatch, the young feed on the body of the larva.

Above: Insects have developed a variety of ways of ensuring that males and females find each other when they wish to mate. An especially ingenious method is used by glow worms. The female emits a strong glow at night which attracts not only our attention but also that of the male beetle.

Below: Many insects, such as the butterfly, have four very different stages in their life cycles. The adult butterfly lays eggs, from which hatch caterpillars. These are the feeding stage in the life cycle. The caterpillars turn into pupae by a process known as metamorphosis, and eventually the adult butterfly emerges, completing the life cycle. More primitive insects have a three-stage life cycle in which the eggs hatch directly into the nymphs, which are miniature versions of the adult.

less detail than our eyes, they give better all-round vision. Insects can also detect smells and temperature and pressure changes.

Social Insects

Above: Bees chew bark to make papery nests. These are sometimes built in crevices or suspended from branches. Workers collect pollen to make wax chambers for their young, and nectar to make honey.

Most insects lead independent lives, coming together with others of their kind only for mating. However, some insects can only survive in complex societies. On their own they would soon die.

In insect societies the whole colony is often based on one breeding female, called a queen. In most cases, all the other members of the society are her offspring. There is usually a well-defined division of labour, with different types of colony members produced for different tasks. This is called a caste system. For example, some members of the colony will defend it, and they have large, aggressive mouthparts. Others may be concerned with collection of food and repairing the nest or hive, and may have chewing mouthparts.

In the ants and termites, the solitary queen is often little more than an egg-laying machine. She devotes all her life to laying eggs but cannot defend or feed herself.

Ants

Ants are interesting social insects which live in large colonies either underground or in loose mounds. Their homes contain a complex system of tunnels and galleries. Ant societies contain three castes: the males, fertile females which become queens, and workers. In the summer, many ants produce wings and fly about looking for mates. These flights are called swarming flights. After mating the fertilized female sheds her wings and searches for a site to start a new colony.

Below: Some ants build large mounds of twigs and other plant material. These contain runs and galleries in which the ants live. Many species of ants have a special relationship with aphids. They collect honeydew from the aphids and in return protect them from predators.

Honey Bees

Honey bees are social insects which man uses to his advantage. In the wild, honey bees nest in cavities, but they take readily to artificial hives. Here their honey can be collected without harming the bees.

There are three castes in a honey bee society. At the centre of the colony is the large queen. She is fertile and lays eggs throughout her life, although she only mates once. The next caste is the drones. These are male bees. Their function is to mate with the queen, and they take no part in the running of the hive. The third caste is the workers, which make up most of the bees in the hive. Like the queen, they are female, but they are sterile. As their name suggests, they do most of the work, gathering pollen, making honey and defending the hive with their stings.

Honey bees show many interesting types of behaviour. In hot weather, they fan the entrance to the hive with their wings to produce a draught. The workers also perform elaborate 'dances' which tell other hive members the directions to the best food.

Termites

Termite mounds, like the ones shown at the bottom of this page, are a conspicuous feature of the African Plains, and are common throughout the tropics. These impressive structures may be up to 6 metres tall. Many are made of cemented soil and so they are virtually impregnable.

Termite society is rather different from those of ants, bees and wasps. There are four castes rather than three. A colony is started after winged males and females swarm. When a pair have settled, they shed their wings and the female becomes the queen of the new colony. She becomes grotesquely enlarged and lays more than 10,000 eggs a day when she is mature. The male is also important because he has to fertilize the queen several times throughout her life.

The soldiers are armed with a huge set of jaws with which to defend the mound from attackers. The fourth caste is the workers, who gather the wood which is the main food for termites. They have special bacteria in their intestines which help them digest it.

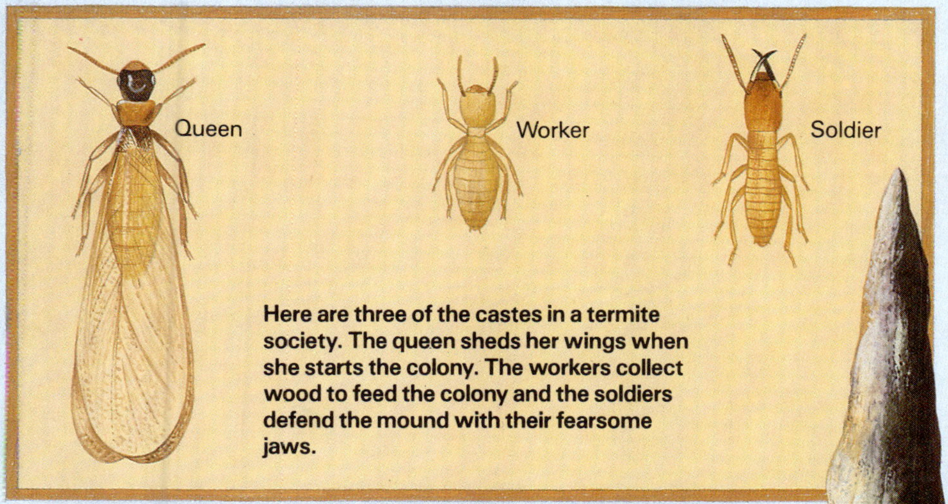

Here are three of the castes in a termite society. The queen sheds her wings when she starts the colony. The workers collect wood to feed the colony and the soldiers defend the mound with their fearsome jaws.

Below: The inset shows the swollen body of the mature termite queen. She grows to an enormous size and lays more than a million eggs in a year. She cannot feed or fend for herself and is looked after by the rest of the colony.

Pondlife

Freshwater ponds often teem with life. Colourful displays of flowers grow around their edges, and a wealth of animal life lives below the surface. The animals which dwell there must cope with very different problems from those of their relatives on land. They must be able to move about in the water and find a way of obtaining oxygen for breathing. The pond dwellers also have to cope with extremes of temperature; sunny days can heat the water up, while in winter the water's surface may freeze.

Some of the pond's inhabitants, such as fishes, can only live in water, but others, such as frogs and dragonflies, spend the early part of their lives in water and their adult stages on land, returning to the pond to breed in the spring.

Life in Ponds

The pond is not a simple environment, but is really a mixture of several separate habitats. The aquatic plants and animals have all evolved to live in very precise zones of the pond. Some plants will grow only in the drying mud on the margin of the pond while others, which cannot survive drying out, must be constantly immersed in water. Duckweed likes to drift in the open water while other species, such as water lilies, remain rooted in the mud at the bottom.

It is, however, the animal inhabitants which are most specially adapted to life in the pond. The surface of the water is very important. It is here that oxygen enters the pond from the air. It is the place where some animals move from air to water, and also where light penetrates the waters below. However, it can also be a prison to many land animals which get trapped in its surface film. Not surprisingly, a variety of scavengers live on the surface and pick off the unfortunate animals caught there. Still more scavengers live underneath this surface film, often trapping air on their bodies to keep them buoyant.

Below the surface is the open water, inhabited by newts, fishes and water beetles. These creatures

Right: This panorama of a pond illustrates some of the creatures which live there. In the water itself, great crested newts and sticklebacks feed on the wealth of insect life present. These in turn sometimes fall victim to the carnivorous great diving beetle. Beneath the surface of the water, many plants grow, rooted in the muddy bottom. When they flower, they either produce floating flowers, such as those of the water lily, or tall flowering spikes like those of reeds and bulrushes. Many birds are associated with fresh water. Herons patiently stalk fishes in the shallows while coots dive for plant and animal food.

feed on others which hide among the tangled pond weeds, and seldom visit either the surface or the bottom.

On the bottom, where the silt collects and where the remains of dead plants and animals fall, many species such as bloodworms and crustaceans tolerate its inhospitable conditions and find it a safe sanctuary away from the eyes of hungry predators.

In order to obtain the oxygen they require for breathing, many freshwater insects come to the surface to breathe, although some, such as damselfly nymphs, have developed simple gills to help them extract oxygen from the water. Many diving beetles have an ingenious way of breathing. They trap a bubble of air under their wingcases and this acts as a kind of 'aqualung'. But, like its human equivalent, it must be topped up from time to time, so the beetle makes regular visits to the surface.

Above: The development of tadpoles into frogs is a remarkable process. In early spring, frogs gather in large numbers in ponds to reproduce. The eggs are laid in great masses, and are surrounded by protective jelly which swells in the water producing the familiar 'frogspawn'. At first the little embryo is just a black dot, but it soon grows to form a tadpole and hatches after a couple of weeks. At this stage the tadpole breathes using feathery gills and rasps algae from the surface of pondweeds. Over the next few weeks it loses its external gills, develops legs and finally loses its tail. At this stage its metamorphosis is complete and the young froglet leaves the pond.

Water plants produce oxygen when they photosynthesize. One little animal named Hydra has found an ingenious way of making use of this. It incorporates small, single-celled green algae between its cells and then 'breathes' the oxygen they produce.

HALFWAY ANIMALS

Frogs, toads and newts belong to a group of animals known as amphibians. They are thought to have evolved from fish ancestors and to have given rise to reptiles. Fishes are almost exclusively aquatic animals, whereas reptiles are almost entirely land dwellers. Amphibians are usually just as much at home in water as they are on land.

When adults, frogs, toads and newts can breathe by taking in oxygen through their skins, but they can also gulp air through their mouths. This enables them to leave the water when they wish. Although some amphibians may spend most of their adult lives away from water, they must all return to the water to breed.

The Seashore

A great variety of interesting and unusual wildlife is to be found on the seashore. In fact, some of the plants and animals of the seashore — such as seaweeds and sea urchins — are found nowhere else.

All the animals and plants of the seashore are extremely hardy, for as well as having to cope with the pounding of the waves they must be able to survive when the tide goes out, leaving them exposed to the drying sun and wind. At this time they are also easy prey for predators, and so must find ways of protecting themselves.

The most common plants of the seashore are seaweeds. Seaweeds belong to a group of plants called algae. The seaweeds grow on the beach according to their ability to withstand exposure when the tide is out.

The most curious seashore plants are the lichens. Some resemble patches of tar stuck to the rocks. Others look like little orange crusty patches.

Types of Seashore

There are several different kinds of seashore. Rocky shores are usually richest in wildlife. The rocks provide good anchorage for the seaweeds, and the crevices offer a safe haven for snails, worms, crabs and other creatures. Many of the snails' relatives, such as limpets and mussels, clamp themselves to the rocks. Countless barnacles, related to crabs and prawns, also cover many of the rocks.

In the rockpools starfishes, anemones and fishes lurk, safe from the wind and sun.

Sandy shores are more difficult places for wildlife to survive. There is little shelter for animals, and so they must burrow beneath the sand to find a moist, secure place to live. Lugworms, fanworms and a great variety of bivalves such as cockles, razorshells and tellins are to be found living in the sand.

Even more harsh environments are shingle beaches. The relentless, grinding action of the pebbles would crush most animals. Therefore the only living things here are specially adapted flowering plants like sea kale.

Above: You will often find shells such as these on beaches. These shells once belonged to animals called molluscs, one of the commonest groups of creatures on the seashore. Other common molluscs include periwinkles and limpets.

Below: Seashores, particularly sheltered rocky shores, teem with animal and plant life. Some creatures live attached to the rocks themselves, while others seek refuge in rockpools or under the sand. Birds such as gulls and waders patrol the shore, looking for food.

ADAPTING TO LIFE ON THE SEASHORE

Some of the seashore animals have strong shells into which they can retreat at low tide, remaining moist and secure until the tide returns. Others are masters of camouflage; certain crabs and worms look like the seaweed in which they hide. The fronds of the seaweed also help to keep the creatures moist.

The seaweeds attach themselves to rocks by means of a root-like structure called a holdfast. They also have soft, slippery stems and fronds which enable them to sway in the currents and avoid becoming damaged. Some species, like bladderwrack, have special air bubbles in their fronds, which help them to float.

Life in the Seas

The seas and oceans cover nearly three-quarters of the world's surface, and are extremely rich and varied habitats. From the shallowest coastal waters to the huge expanses of the open ocean, and from the icy polar waters to the tropical seas, they literally teem with life.

Shallow Seas

In the shallow coastal waters we often find plants and animals similar to those found on the seashore, such as seaweeds, crabs, snails, bivalves (molluscs with hinged shells), starfishes and many other familiar creatures such as seals and sea lions.

In the warmer parts of the world the shallow waters are often the home of corals. These tiny creatures, related to anemones, build chalky homes around themselves. Eventually these may extend for many kilometres. Coral reefs are fascinating places in which fishes, starfishes, sea urchins and anemones and other creatures all live together in a bewildering array of colours and shapes.

Above: These dazzling clown fish are among the many colourful species which inhabit coral reefs. The world's biggest coral reef is the Great Barrier Reef, extending for 2012 kilometres along the north-eastern coast of Australia.

Below: Most sharks are fearsome predators of the sea, although a few species eat only plankton. Dolphins and porpoises are kinds of whale. Whales are mammals which spend all their lives in the sea. Many seabirds feed on fishes which they catch near the surface.

Open Oceans

The smallest forms of life in the open sea are the plankton. These are tiny, microscopic plants and animals which float near the surface. They provide food for many of the other species of animal living in the sea. Even some of the huge whales and sharks live just on a diet of plankton. Most of the sharks, however, are fierce predators of other animals, particularly the huge shoals of fishes which roam the oceans. Species such as the hammerhead and the great white shark are even feared by man, for attacks on humans are not uncommon in waters in which they are found.

Turtles and sea snakes are reptiles frequently encountered in the open sea. Turtles are generally found in warmer waters. They spend most of their lives at sea, only hauling themselves on to beaches at night to lay their eggs in the sand. As soon as they hatch, the baby turtles must scuttle back down the beach to return to the safety of the sea.

Sea snakes are highly venomous, and about 50 species are known to inhabit the sea, mainly in the waters of the Indo-Pacific region. They feed on small fishes.

Deep Sea

In the gloom of the deep sea, where light never reaches, some of the most bizarre creatures are to be found. These include giant squids, prawns and angler fishes.

Above: Jellyfishes and the Portuguese man o'war drift in the open oceans. They catch fishes and other small creatures, using their trailing tentacles which are armed with powerful stinging cells.

Below: The deep seas of the world are inhabited by species like the ones shown here. Many deep-sea creatures have special light-producing organs on their bodies to attract small prey.

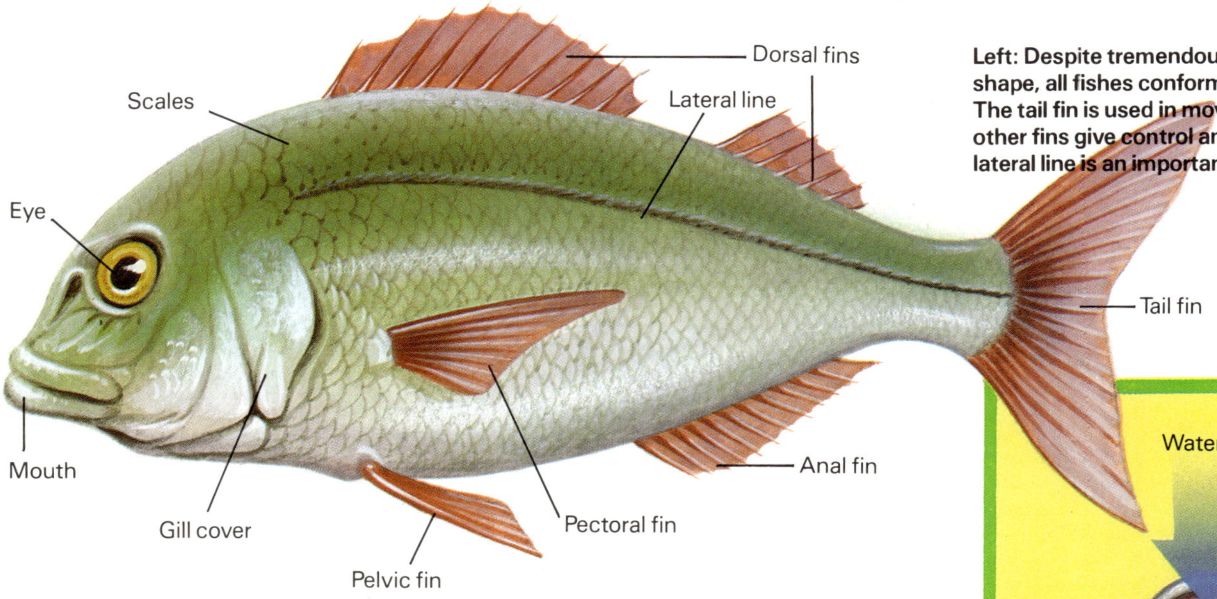

Left: Despite tremendous variations in body shape, all fishes conform to a basic plan. The tail fin is used in movement and the other fins give control and balance. The lateral line is an important sense organ.

All Kinds of Fishes

The fishes are a very varied and successful group of vertebrate animals which have colonized the world's oceans, lakes and rivers. Fishes are cold-blooded creatures whose bodies are covered with scales. They are specially adapted to live and breathe in water, although some can spend limited amounts of time on dry land.

Among the most primitive of fishes are the lampreys and hagfishes. These eel-like creatures do not possess proper bones nor a proper set of jaws. Nevertheless, they are a successful group and have, instead of jaws, a rasping sucker. They feed by attaching themselves to the sides of other fishes and rasping away at their flesh.

Sharks, rays and skates lack proper bones as well, but they have a strong cartilage skeleton to support their bodies. Most sharks have a fearsome array of teeth in their powerful jaws. They are mainly predatory creatures which hunt other fishes as well as squid and various invertebrates. In tropical waters there are species of shark which reach huge sizes. Both the feared tiger shark and the man-eating great white shark often exceed 6 metres in length. The massive whale shark, on the other hand, is harmless to man, feeding on microscopic plankton. Skates and rays are relatives of sharks. Their bodies are flattened, for they spend much of their life on the seabed.

The most numerous group of fishes are the bony fishes. Most of those which we eat, or keep as pets, are in this group. The extraordinary lungfishes are found in tropical lakes. They survive the dry season by curling up in a mud chamber. Their 'lung' allows them to breathe air through a small opening in the chamber.

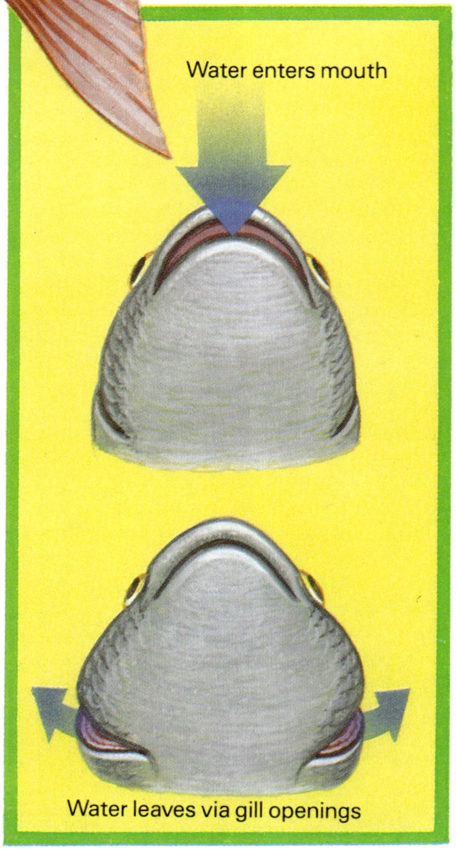

Above: Fishes breathe by extracting dissolved oxygen from the water, using a series of internal gills over which water is continually passed. The process can easily be observed in aquarium fishes and first involves the fish taking in water through the mouth. With the mouth closed, a swallowing movement forces the water over the gills and out through the gill openings.

Right: The pike is often called the freshwater shark, because of its voracious appetite and its fearsome array of teeth. It takes its prey by surprise and its body is well adapted for this purpose. The markings on its flanks provide camouflage as it lurks in the weeds, and its large tail fin gives it excellent acceleration. With its huge gape it can easily grab prey and the teeth prevent its escape.

Other Bony Fishes

The cod family are large, bony marine fishes which provide food for man as well as many other fishes and birds. They usually have large eyes and mouths, and powerful bodies which enable them to swim well. Some of them can live in very deep water. The herring family are also important for food, and are found in vast shoals in some seas where many birds and other creatures depend on them for food. The salmon family are streamlined and powerful swimmers, and many of them live in the sea, but make long migrations up rivers to spawn.

In fresh water, members of the carp family are very common, ranging in size from tiny minnows to the large carp themselves, and they include the familiar goldfish.

Tropical lakes and rivers support large numbers of fishes from the cichlid family which are often brightly coloured and very active swimmers.

The perch family are armed with spines and strong scales and hunt other fishes for their food.

Above: When the young plaice hatches from its egg it is much like any other bony fish. After a few days, however, changes begin to take place and one eye moves across the top of its head. Eventually it ends up alongside the other and the young fish begins to swim on one side with its eyes uppermost. It is now perfectly adapted to life on the sea bottom.

Far left: In the spring the male stickleback builds a little nest into which he lures a female. If satisfied with the arrangements, she lays her eggs in the nest and departs. For a couple of weeks, the male guards the eggs and looks after the young when they hatch, herding them back into the nest at the first sign of danger.

Above left: The mouthbrooder protects its eggs in a most unusual way. After they have been fertilized, the female collects the eggs in her mouth and carries them around until they hatch.

The Reptilian World

Reptiles are scaly, cold-blooded vertebrate animals which lay eggs with hard, protective shells. The reptiles descended from amphibians 300 million years ago. Today, reptiles are found mainly in the warmer parts of the world.

Many millions of years ago, there were many more kinds of reptiles on the earth than there are today. During the period in the earth's history known as the Mesozoic Era, huge reptiles roamed over many continents. Other reptiles conquered the sea and air. About 70 million years ago many reptiles died out, however.

Above: The anaconda is a large snake from South America which kills its prey by constriction or squeezing, before swallowing it whole. The anaconda is also found in rivers.

Left: A cobra rears up in this position before striking at its prey.

Below: Snakes can open their mouths extremely wide. This is how they are able to swallow big prey.

Living Reptiles

The biggest group of living reptiles is the Squamata. These include the snakes and lizards. Snakes are reptiles which have lost the use of their legs over the course of evolution. They move along by means of special movements of their muscular bodies, and they also use their scales to help obtain a grip as they move. A few species swim in the sea.

All snakes are predators, feeding on many different kinds of animal food from eggs to creatures as big as pigs. Some snakes kill their prey before eating it by first squeezing it to death. To do this they quickly surprise their victim and wrap their coils tightly around it. Smaller prey is often swallowed alive. Snakes such as adders, cobras and rattlesnakes first inject their prey with a venom before swallowing it.

Lizards have adapted to many different ways of life. They are also predators but, unlike the snakes, only two species are venomous. The biggest lizard is the Komodo dragon of Borneo. This huge creature measures 3 metres in length. It eats large prey. Most lizards, however, feed on small mammals and birds, as well as on insects and other invertebrates.

The Crocodilia forms the second group of living reptiles. This group includes the crocodiles, caimans, alligators and the gharial. Members of the Crocodilia look very much like their ancestors which existed when reptiles ruled the earth.

Right: The chameleon is a curious lizard which creeps about branches looking for insects to eat. It can swivel both its eyes in different directions. It catches its food on the end of a long, sticky tongue.

Below: The heads of three different crocodilians.

Crocodilians look like huge, armour-plated lizards. They are all adapted to a life spent in water, where they lurk partly submerged beneath the surface waiting for prey. Sometimes they will snatch animals that come to the water's edge to drink. At night they may leave the water to sleep on a sandbank.

The tails of crocodilians are specially flattened for swimming, and the nostrils are placed high on the snout, enabling them to breathe even when lying partly submerged in water.

Crocodilians are found in many parts of the world. They live in the fresh waters of tropical Asia, Africa, Australia and America. One species, the estuarine crocodile, is found in the estuaries of rivers in parts of Asia and Australia. It also swims out to sea.

The most unusual-looking crocodilian is the gharial of India. It has long, thin jaws armed with sharp teeth. Despite its fearsome appearance, however, it catches nothing bigger than fishes.

The Chelonia comprise the tortoises, terrapins and turtles. The chelonians carry on their back a bony shield, and their underbodies are protected by a large, flat, bony plate. When danger threatens these animals can pull their heads and limbs inside their protective armour. Unlike other reptiles, chelonians do not possess teeth, but they have sharp, bony beaks. Also, unlike other reptiles, some chelonians are herbivorous.

Although there are exceptions, most tortoises are land-living creatures, most terrapins live in fresh water, and turtles live in the sea.

The last group of living reptiles is the Rhynchocephalia. Only one living member of this group exists. This animal is called the tuatara. It lives on remote islands near New Zealand, and looks rather like a lizard with spines on its back. It often shares a burrow with a bird, the Manx shearwater.

Below: The giant tortoise of the Galapagos Islands is big enough for a man to ride. A common tortoise is shown by its side.

Birdlife

It would be impossible to mistake a bird for any other animal, for only birds possess feathers. It is feathers which have given birds mastery of the air.

There are nearly 9,000 species of birds throughout the world today, living in every sort of habitat from the seashore to the jungle. The smallest birds are the brilliantly coloured hummingbirds, whose nests are no bigger than a thimble. At the other end of the scale is the ostrich, a flightless bird standing 2.5 metres high.

Birds have adapted to feed on all manner of food. There are some species, the birds of prey, which hunt other animals. Others eat seeds and fruit. A few feed on nothing but insects. Hummingbirds hover over flowers and suck up nectar.

Above: The eagle has a sharp, flesh-tearing beak.

Above: Ducks have wide, flat beaks for sifting mud.

The Development Inside a Bird's Egg

Bird Flight

The ability to fly is the key to the great success of birds. Flight has enabled birds to travel huge distances in search of new territories; to travel from continent to continent during migration; and it has allowed them to escape from their enemies when danger threatens.

For successful flight birds must have lightweight, streamlined bodies. Bird bones are hollow, but specially strengthened, and the feathers help to give them a smooth shape in the air. Lift and motion through the air is provided by the feathers on the wings. The tail is used for steering. Birds which spend hours soaring over oceans have long, narrow wings. Birds which fly fast have sickle-shaped wings. Birds which must fly between the trees have shorter, powerful wings.

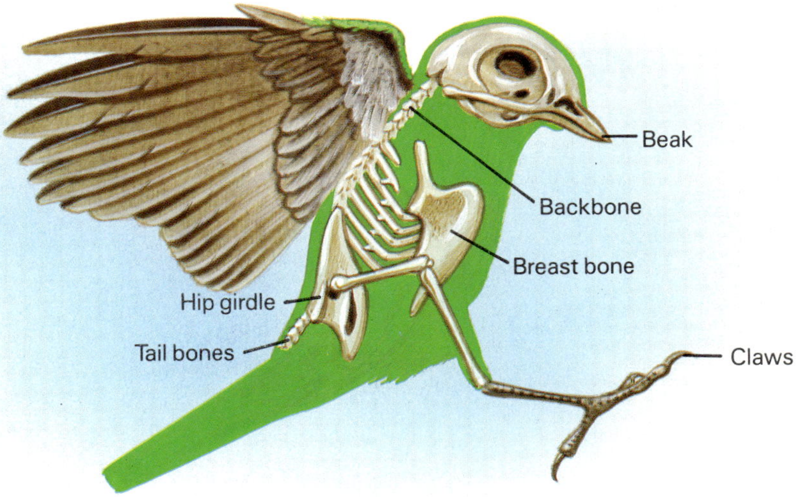

Left: This is how the skeleton is arranged inside a typical bird such as a crow.

Below: Crossbills have short, stout beaks for eating seeds.

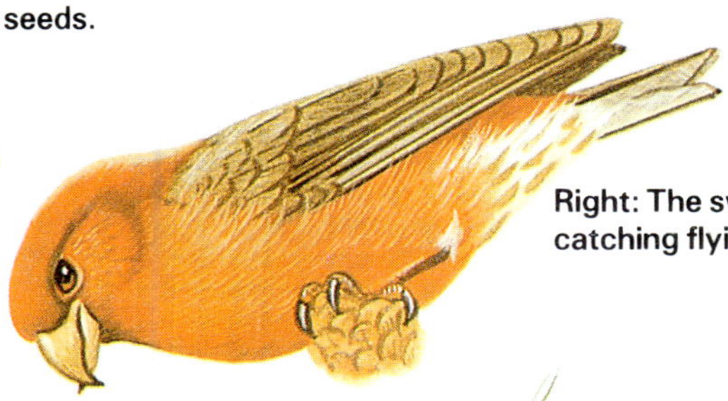

Right: The swift has a very wide beak for catching flying insects.

Above: Skimmers use their beaks to catch fishes near the surface.

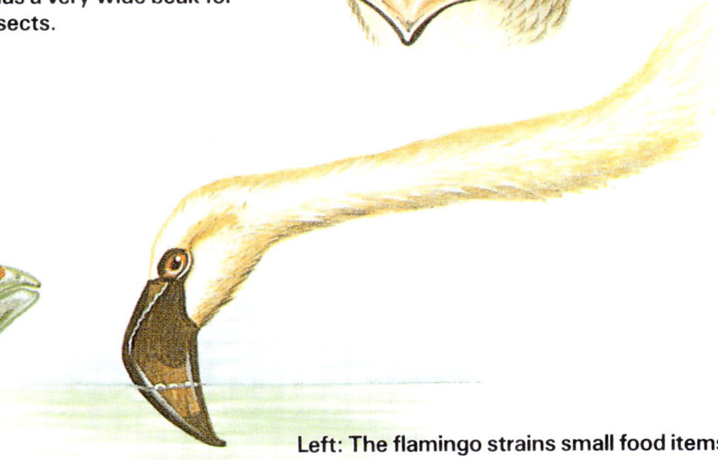

Left: The flamingo strains small food items from the water.

Birds and their Young

Many animals build a nest in which to rear their young in safety, but the most familiar of these are birds' nests. Not all birds build nests, however. Some species simply lay their eggs on the ground. The cuckoo is an unusual bird which lays its eggs in the nest of another species.

Before nesting begins, most birds choose a territory and defend it against rivals. The number of eggs laid by the female depends on the species. It may be just one, or as many as 20. Once hatched, the young chicks are fed by their parents until they are big enough to fend for themselves. Most chicks are blind and helpless at birth, but the chicks of waterbirds are born fully feathered and can swim straight away.

Great crested grebes have an elaborate courtship display known as a mating dance.

Right: Many birds, like this whip-poor-will, use their plumage for concealment.

Birds Around the World

Birds are found all around the world, from the hottest deserts to the icy polar caps. Some are specialized and have restricted distributions. Thus, secretary birds are found only on the African Plains, and hummingbirds are found only in the Americas. Other species have wider distributions. The golden eagle, for instance, is found all over the northern hemisphere.

Birds are grouped into families which share recognizable features. Some families have representatives in all parts of the world. For example, ducks, geese and swans can be found in almost every habitat, and all share a similar appearance with webbed feet and a flattened bill. Puffins, and their relatives the auks, on the other hand, are only found in the northern hemisphere.

Nectar Eaters

Hummingbirds are a very specialized group of birds which can hover while collecting nectar from flowers. They are usually very colourful birds with iridescent plumage. Hummingbirds are only found in the Americas, with most species occurring in Central and South America.

Bird Movements

Despite their powers of flight, some birds move very little during their lives. Ptarmigan, for example, are arctic gamebirds which remain in the same habitat even under the harshest of winter conditions, and seldom fly unless disturbed. In complete contrast, albatrosses and shearwaters spend the major part of their lives at sea on the wing, ranging across the oceans of the southern hemisphere. They spend only a few short weeks ashore each year during the breeding season.

In Africa, the sunbirds have also evolved to collect nectar. They cannot hover and are not related to hummingbirds, but look remarkably similar at first glance.

Birds of Prey

Birds of prey are found all over the world. They have sharp talons which they use to grip their prey, and a hooked beak to tear the flesh. They range in size from the tiny pygmy falcon, no bigger than a sparrow, to immense eagles which can soar effortlessly for hours on end.

Some birds of prey are specialized to feed on carrion. These are the vultures and the condors. They are not very closely related, although they look similar. Condors are found only in the Americas whereas vultures only occur in Europe, Africa and Asia.

Birds of the Antarctic

The most characteristic birds of the antarctic are the penguins. These rather comical birds feed on the abundant marine life in the polar seas.

There are many different species but they all share the same characteristics. They have lost the ability to fly. Instead, they use their wings as flippers to swim, almost like flying, under water. Their feathers form a dense, insulating layer against the freezing waters and biting wind. They are an extremely successful family, and colonies often number over a million birds.

Below: Birds are found on all continents of the world.

Apes and Monkeys

Apes and monkeys, together with the tree shrews, lemurs, lorises and bush-babies, make up the group of animals known as the primates. In many ways primates are extremely interesting creatures, because man is also a primate. Many monkeys and apes spend their lives in the safety of the treetops. A few, however, have returned to live a life on the ground. Ground-dwelling species include the baboon, the gorilla and, of course, man himself.

Apes and monkeys are intelligent creatures. They have relatively large brains, and forward-facing eyes which help them to judge distances well. Their hands and feet are designed to help them grip branches as they leap from tree to tree. They often live together in well-organized family groups.

These three animals are primitive primates related to the monkeys and apes.

Below: Baboons are large, African, dog-like monkeys which live mainly on the ground. They live in family groups and eat all kinds of food. They defend themselves fiercely if attacked.

Above: Gibbons are apes with very long arms. They swing through the trees with great ease, but can also run along branches. They live in Asian forests and feed mainly on fruit.

Above: Chimpanzees are highly intelligent apes which spend part of their lives in the treetops of African forests and part on the ground. They can use their hands in a way very similar to us.

Right: Spider monkeys are extremely agile climbers which can also use their tails to hang on to branches.

Below left: Mandrills are forest-dwelling baboons which live in Africa.

Above: The orang-utan is an ape which lives in Borneo and Sumatra.

Left: Despite their huge size and fierce appearance, gorillas are shy, plant-eating apes.

Monkeys

Monkeys are divided into two groups: those that live in places like South America, and those that live in Africa, India and Asia. Some monkeys are active during the day, where they search for fruit, leaves, insects or other small animals to eat. Others come out only at night, when their strange calls fill the warm night air.

Apes

There are ten species of apes, all of which live in the tropical forests of Africa and Asia. One group, which includes the gorilla, orang-utan and chimpanzee, is known as the ‘great apes’. Apes are the group of primates most closely related to man.

Cold Lands

The polar regions are the coldest and bleakest places on Earth. Antarctica, around the South Pole, is mostly covered by thick ice, and fierce blizzards blow loose snow over the surface. Temperatures rarely rise above freezing point during the day and a world record −88.3°C has been recorded near the South Pole. No people live permanently in Antarctica, but some scientists work there in heated homes under the ice. Penguins live around the shores of Antarctica and the seas are rich in fish and whales.

Polar bears live on the ice floes of the Arctic Ocean, hunting seals and young walruses. And around the Arctic Ocean are parts of North America, Europe and Asia, called the *tundra*. Here the snow melts during the short summers, when temperatures may rise to 10°C. Flowering plants then carpet the ground, but it is too cold for trees. Migrating animals such as reindeer graze in the tundra. Arctic people include the Eskimos. They once lived by hunting. They built winter homes of ice called igloos, but most Eskimos now have modern homes.

Arctic

Antarctica

Above: One map shows the ice-covered continent of Antarctica around the South Pole. The other map shows the region around the North Pole, which contains the icy Arctic Ocean. The northern parts of North America, Europe and Asia surround this ocean. These lands have long, cold winters, but plants grow in the short summers.

Below: Penguins gather in large colonies on the icy coasts of Antarctica. The emperor penguin hatches its egg on its feet and the young birds huddle together for warmth.

Most people in the great Sahara Desert live around oases. These fertile places have wells or springs which tap water in rocks below the surface. Some nomads cross hot deserts in camel caravans. Camels are called the 'ships of the desert'. This is because they can go long distances without drinking. But when they reach an oasis, they often drink up to 100 litres of water in only 10 minutes. The camel is suited to the desert in another way. A web of strong skin connects its two widely spaced toes and prevents its feet from sinking into the sand.

Many animals live in the desert despite the heat and the dryness. The smaller animals hide away during the day to escape the fierce sun and venture out at night. Many of them get the water they need from the seeds they eat.

Hot Deserts

Hot deserts cover about one-fifth of the world's land areas. They have mostly clear skies and high temperatures. The average rainfall is under 250 millimetres a year. Several years may pass with hardly any rain, then a thunderstorm may drench the land. Seeds which have been dormant for years rapidly germinate and many plants flower. When this happens in the Australian desert, people flock to the area to see this superb spectacle.

The world's largest desert is the Sahara in North Africa. It contains sand dunes, plains of loose gravel and bare, rocky uplands. Most plants in hot deserts, such as thorn bushes and cacti, are drought-resistant. The most useful desert animal is the camel. It can go for long periods without water. It loses weight, but regains it when it drinks. Deserts become fertile when they are irrigated (watered). Some deserts, like those of the Middle East, are rich in oil and natural gas. These fuels lie hidden far below the surface.

Penguins are awkward creatures on land, waddling around on their webbed feet. To move faster they slide on their bellies over the ice, pushing themselves along with their stubby wings. In the water penguins are fast and agile as they chase fish under water.

Grasslands

Tropical grasslands, called savanna, campos or llanos, occur north and south of the equator in Africa, South America, southern Asia and northern Australia. They are in warm regions with high rainfall. But, because they have a marked dry season, forests do not develop except in such areas as river valleys. Tropical grasslands are used for cattle rearing and some crop farming. But there is a danger that the winds will blow away the exposed soil in the dry season. Such soil erosion can make a region infertile in a few years. In Africa, the savanna supports many animals.

Temperate or mid-latitude grasslands have less rain and much colder winters than tropical grasslands. Trees are also rare on the dry, windswept plains. Huge areas of temperate grasslands, which are also called prairies, steppes or pampas, have become cattle ranches or vast wheat farms. The wildlife has been much reduced. In Australia, sheep have replaced most of the kangaroos and wallabies of the grasslands.

Above: Cattle ranching is a major activity in such savanna regions as East Africa.

Below: Temperate grasslands in North America have become leading wheat-growing regions. The land is flat and easy to farm.

Woods and Meadows

Above: Forests once covered the moist temperate regions in the middle latitudes. In England, these forests have been replaced by a complicated patchwork of fields.

Below: Moist temperate regions in New Zealand support great numbers of sheep.

Moist temperate regions have some rain all the year round, with warm summers and fairly cold winters. Such regions once had huge forests of deciduous trees, such as ash, beech, chestnut, hickory, maple and oak. These trees shed their leaves in winter in order to protect themselves against the cold. Such forests once grew over most of western Europe, northern China and the eastern USA, areas which now contain some of the world's most densely populated regions. Most of these forests have been cut down. Their wood was used for building and as a fuel. In the USA, the forests largely vanished in about 300 years and the land was used for farming. The wildlife dwindled and soil erosion became a serious problem. This also happened on the lower slopes of the mountains of South Island, New Zealand. Britain's forests were replaced by a pleasant landscape of ploughed fields, river meadows, hedgerows and occasional clumps of woodland. But the destruction of forests led to a great decline in wildlife.

Northern Forests

Cold snowy climates are warmer than polar regions, but colder than temperate regions. Vast coniferous forests, called taiga, grow in cold snowy climates. Coniferous trees include birch, fir, pine and spruce. These cold forests are found only in the Northern Hemisphere. No large land masses in a similar latitude exist in the Southern Hemisphere.

Many furred animals, such as ermine, mink, otter and wolverine live in the forests. They have attracted hunters, who have greatly reduced the numbers of these animals. Larger animals include bears, bull moose, caribou, reindeer and wolves. Many trees in these cold forests are valuable softwoods. They are cut down and floated along rivers to sawmills. Sawn timber is used to make furniture and many other things. Many logs are made into wood pulp and paper. The northern forests do not contain many people, because of the cold weather and their generally infertile soils.

Above: The map shows the vast belt of cold coniferous forest that sprawls over North America, northern Europe and northern Asia. It lies south of the tundra and north of the deciduous forests and grassland.

River Valleys

Rivers carry worn pieces of rock from their upper courses to the sea. As the rocks move along, they break down into smaller and smaller pieces. Finally, they become fine alluvium. This alluvium is spread over river valleys when rivers flood. It sometimes piles up in deltas, areas of new fertile land at the river mouths.

Fertile river valleys and deltas are the most densely populated regions on Earth. Farmers can produce large crops because of the rich soil and plentiful water. Some valleys have been centres of early civilizations. These include the valleys of the Tigris-Euphrates in Iraq, the Nile in Egypt, the Indus in Pakistan, the Ganges in India, and the various river valleys of eastern China.

Above: Rivers flow swiftly down their steep upper courses but they become slow-moving when they near the sea. The lower valleys and deltas of many rivers are thickly populated. In parts of China, including Hong Kong, people live on river boats called sampans, right.

Below: Buffalo plough the flooded 'paddy' fields in which rice is grown. Rice from the great river valleys is the staple diet of Asia.

Animals of Australia

Australia is a huge continent in the South Pacific. Together with other islands such as New Zealand and New Guinea, it forms an area called Australasia. Naturalists have always been fascinated by the wildlife of Australia, for many of the animals there are found nowhere else in the world today.

Many millions of years ago the land masses of the Australasian region were connected to the rest of the world. All the animals roamed freely across the continents because at that time they were not separated by great oceans.

Then, the continents began to drift apart. Australia and the other countries of Australasia slowly moved towards the South Pole. Now, all the kinds of animals which were living there became separated, or isolated, from all the other countries of the world by thousands of kilometres of ocean.

Before the drift began most of the mammals were of the kind called marsupials. They had pouches in which the young developed. There were also some other, even stranger, mammals called monotremes. They laid eggs. Cut off from the rest of the world, Australia's mammals developed into many species quite different from the mammals which later arose in other parts of the world.

Above: Some of Australia's birdlife is also very unusual. The kookaburra is related to the kingfishers. It has a very distinctive call and because of this is also known as the 'laughing jackass'. Some Australian birds, such as the bee-eater, birds of Paradise, parrots and cockatoos, are beautiful birds with extremely colourful plumage. Birds of Paradise and parrots live in dense forests, but bee-eaters prefer sandy banks and kookaburras live near water.

INTRODUCED MAMMALS
The arrival of the first men in Australia several thousand years ago caused some important changes. The earliest men brought with them domestic dogs. Some of these returned to the wild and began to prey on the marsupial mammals, which were unable to defend themselves. The later settlers brought sheep and cattle. The farmers who tended these animals drove many of the native animals off their grazing land. Other introduced mammals such as rats and rabbits also competed with the marsupials for food and territory. Today, many marsupials are in danger of extinction, and special measures have been taken to protect them.

The platypus is one of only two mammals in the world which lay eggs. The other egg-laying mammal is the echidna, or spiny anteater. Both of these mammals are known as monotremes. Although these mammals lay eggs, they still feed their young on milk just as other mammals do. Monotremes also have a beak, instead of teeth, and the males have a poisonous claw on their hind legs. The platypus lives in burrows in river banks. It spends most of its life in the water, where it swims about searching for small creatures to eat.

Right: This scene shows some more Australian animals. Many of Australia's animals live in the hot dry interior of the country known as the outback. The frilled lizard raises the frill around its neck in order to frighten would-be predators, although the animal is in fact harmless. Note the young kangaroos, known as joeys, which are carried in their mothers' pouches. Flocks of budgerigars roam across Australia, feeding on seeds. In some places they are considered pests, as they feed on crops. In the forests of eastern Australia the koala is to be found. This mammal looks like a small bear with tufted ears. It feeds on eucalyptus leaves and is only active at night. The emu is a large, flightless bird. Although it cannot fly, it can run extremely fast to escape from its enemies.

Life on the African Plains

Below: A cheetah chases after a group of impala, scattering giraffes and ostriches as it does so. Although very young elephants might fall victim to predatory cats, the cheetah would not dare attack with a protective adult elephant close by.

The African Plains are vast belts of grassland stretching across the continent. Some of the African Plains occur in the temperate regions of southern Africa, but most are found in the tropics. Rainfall occurs for only a short period in tropical grasslands. When it rains the grass grows high and lush. Between the rains the land looks rather barren, with few trees.

The grasses provide food for a variety of large animals such as elephants, giraffes, deer and zebra. Smaller animals like insects and snakes find shelter and food among the grasses, too.

Life on the African Plains can be harsh. The grazing animals must constantly be on the lookout for predators, and they may struggle to survive in years when droughts occur. During droughts, the grasses shrivel and die, fires are common, and even waterholes may dry up. Then, the great herds of grazing animals must often roam far in search of water and food.

Animals of the African Plains

One of the most astonishing sights in nature is to see the huge herds of animals such as wildebeest, many thousands of individuals strong, as they move slowly across the plains. Sometimes herds of different animals such as zebras and antelopes will join together for mutual protection, the whole herd alert to danger.

Few of the grazing animals compete with each other for food, for each type specializes in eating different vegetation. The tall acacia trees can only be reached by giraffes, smaller bushes are eaten by other grazers like eland, and the grasses are eaten by rhinoceroses, zebras and gazelles.

The birdlife of the African Plains includes the huge, flightless ostrich as well as birds of prey like the secretary bird, which attacks and eats snakes. Hidden among the vegetation are hordes of insects, spiders and other small creatures. These provide food for small rodents and reptiles, which themselves may fall prey to other predators.

Predators

The many herds of grazing animals provide food for the carnivorous animals of the African

Plains. The best known of these are the cats—the lions, cheetahs and leopards. Antelopes and zebras form the main food for the cats. Each of the cats has its own special way of catching food. Lions normally hunt in groups. They may lie in wait at a waterhole or surprise a herd on the open plain. Often some of the lions in the group will chase prey to where other lions are waiting.

Cheetahs rely on their great speed to catch prey. They will wait silently, crouched among the grass, until they single out a victim from the herd. Then they bound after their prey at great speed. Cheetahs can run at over 100 kilometres per hour for short distances, but they soon become exhausted and give up the chase if they have not caught their prey within about 40 metres. Leopards eat small antelopes and birds.

Once the prey is brought down and killed with a bite to the neck, it is dragged to safety and eaten. Members of a pride of lions will often feed on the prey together, the strongest individuals taking the best pieces. When the cats have had their fill, it is time for other creatures of the plains to join the feast. First to arrive will often be the hyaenas, which scrap noisily between each other over the left-overs. Then come the vultures, and finally the scavenging insects.

In the Amazon Jungle

The Amazon Jungle is part of a huge rain forest covering much of tropical South America. Rain forests are special kinds of forest. Here, the temperature is high all year round and heavy rainfall occurs frequently. These conditions are just right for rich plant growth, and within rain forests we find many kinds of luxuriant trees and other plants such as orchids, ferns and bromeliads.

LIFE IN THE JUNGLE CANOPY
Although some animals are to be found living on the floor of the rain forest, many more creatures live among the canopy. Brightly coloured birds like toucans and parrots fly among the treetops, and groups of monkeys live in noisy family groups. Here we also find the predatory animals such as snakes, and harpy eagles which swoop down to snatch birds and small monkeys in their sharp talons. Sometimes predatory cats such as jaguars also climb the trees in search of prey.

Right: The world's tropical rain forests are found in equatorial regions where the temperature is always above 65°F and where at least 200 centimetres of rain falls each year.

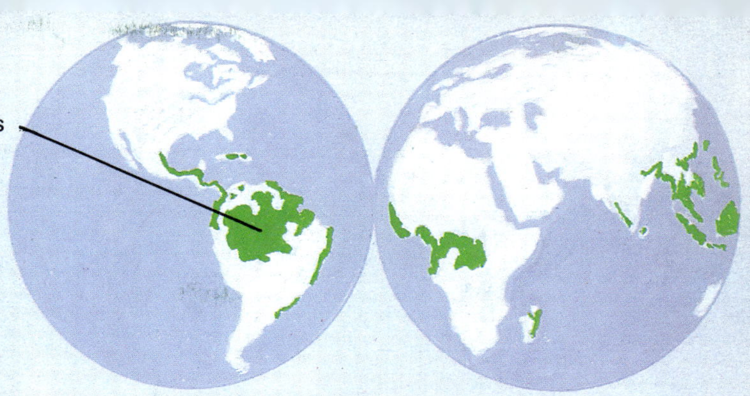

The trees in rain forests consist of species like ebony, teak and mahogany. They grow close together, with their upper branches thrusting towards the sky. The top-most parts of the trees form a layer known as the canopy. The leaves of the canopy layer shade most of the sunlight due to their density, and so the forest floor is gloomy, with little plant growth. Most of the other plants of the rain forest grow on the trunks and stems of the trees, about 15 metres off the ground where some sunlight still penetrates. Plants which grow on other plants are called epiphytes.

Below: Seen from the air, the Amazon rain forest appears as a dense canopy of leaves. These upper branches are the home of many animals, such as monkeys, birds and snakes. On the floor, where little light penetrates, live creatures such as huge beetles and centipedes, as well as deer and the pig-like tapir. Jaguars move stealthily among the tree roots, waiting to pounce on their prey.

Life in the Mojave Desert

Deserts are found in parts of the world where there is very little rainfall. Most deserts are hot, dry places but some are very cold. On average, less than 25 centimetres of rain falls each year in deserts. Sometimes it may not rain in the desert at all for several years. Then torrential rain will fall, bringing floods which sweep through the desert.

Deserts are usually rocky or sandy, with little vegetation. The plants and animals which live in deserts have to find special ways of living in the harsh, dry conditions.

The Mojave Desert is in southern California, in the U.S.A. It covers an area of about 38,850 square kilometres.

Animals of the Mojave Desert

As in other hot deserts, animals of the Mojave Desert have developed special methods to help them keep cool and conserve water. Many animals burrow underground by day. In the Mojave Desert the temperature may be 35°C lower underground than on the surface by day. At night, when the temperature has dropped, the animals leave their burrows to hunt for food.

Animals such as desert rats have large eyes which help them to see at night. Hunters such as snakes rely on sensing the warm bodies of their prey in order to track them down.

When desert animals must travel about during the daytime, they move very fast so that they do not remain too long in contact with the hot ground. The roadrunner uses its great speed to catch lizards and small snakes.

Plants of the Mojave Desert

The largest and most spectacular plants of the Mojave Desert are the cacti. Saguaro cacti can grow to over 16 metres in height. Cacti absorb water very quickly when it rains and then store it in their thick leaves and stems. Plants which can do this are called succulents. The creosote bush is another strange desert plant which can actually survive being dried up by the sun's rays.

When the rains come, all the desert plants quickly bloom, and seeds which had lain dormant in the ground suddenly germinate. For a short while the barren desert becomes a flower-filled landscape.

Above: The desert is full of animal life, although most species spend the day hiding away in the shade. The map at the top shows the main deserts of the world.

Ends of the Earth

The polar regions cover the most northern and most southern areas of the globe. In the northern hemisphere the polar region is called the arctic, and in the southern hemisphere it is called the antarctic. Animals that live in the polar regions have to endure some of the harshest conditions on earth. There are permanent snow and ice fields over much of the areas, and there is always the chance of a sudden icy blizzard, even in summer.

Rich Life in the Seas

Most of the land in the polar regions is either covered with snow, or is frozen just below the surface. As a result, few animals can find food here.

However, the polar seas are extremely rich in food, and provide food for birds such as penguins, and mammals such as bears, whales and seals.

Long Days and Nights

In the polar regions, the sun never climbs very high in the sky, even at the height of summer. As a result, its warming effect is never properly felt. However, throughout the summer it never sets, giving the polar inhabitants perpetual daylight.

In the winter, however, the sun only rises above the horizon each day for a few hours at most. At the poles themselves it does not rise at all for several weeks. Many of the animals move away from the poles in winter to find milder climates and more daylight.

Top left: Some animals stay and endure the arctic winter. A few, like the ptarmigan, turn white to camouflage them against the snow.

Left: In the summer, the arctic tundra is visited by many migrant birds such as waders and geese. The ptarmigan and arctic fox lose their white coloration.

Below: The dramatic antarctic is home to millions of penguins.

How Animals Communicate

Animals communicate with each other for several reasons. Between members of the same species they may communicate in order to find a mate in the breeding season, or to warn other rivals that they have chosen a particular place as their territory. Watch a bird such as a robin. In the breeding season it will fly from tree to tree, or perhaps perch on a fence, and each time it lands it will sing. In fact, what it is really doing is flying around the boundaries of its territory, telling other robins to keep out.

The fiddler crab communicates by means of 'hand signals'. When the male wishes to attract a female, it sits on the sand and beckons the female with a wave of one of its pincers, which is specially enlarged.

Above: Songbirds, like the American robin shown here, sing to warn other birds that they have chosen a territory, and to attract a mate.

Left: Male moths can pick up the scent given off by females using their feathery antennae.

Below: The male fiddler crab tries to attract females to mate with him by waving one of his pincers, which is specially enlarged.

Whales swimming through the vast oceans of the world communicate by sound. Scientists have often recorded the mournful calls of the humpback whale, which can carry over hundreds of kilometres.

Animals often tell each other about the presence of food, or warn each other of possible danger. Sometimes one bird in a feeding flock will alert the others to a source of food and the rest of the flock will quickly gather to feed. In order to survive, different species of animals have learned to heed each other's warnings. The frantic clap of wings made by a pigeon alarmed in a wood is also a signal to other animals such as squirrels and deer that danger is present. And when lions come to feed at an African waterhole the first grazing animal to spot the danger quickly alerts the others, as zebra, wildebeest, gazelles and even elephants all run to safety.

Many animals are poisonous, and this prevents them being eaten

Right: Timber wolves live in forests and snow-covered wastes. They communicate to other members of the pack by means of special calls. This helps to keep the pack together, even at night.

by predators. Animals which are poisonous, or which are highly venomous, advertise this fact by having brightly coloured bodies. Coral snakes, arrow-poison frogs, ladybirds, kingfishers and wasps are just a few of the many animals which tell other animals that they are poisonous. Sometimes other harmless animals mimic these poisonous species in the hope that they, too, will be left alone by predators.

Right: The skins of arrow-poison frogs contain a very powerful poison, and their brightly coloured bodies warn predators that they are poisonous. The African elephant gives warning to its enemies by drawing its ears forward in a threat display. The caterpillar of the Puss moth is harmless, but tries to alarm would-be predators by rearing up and showing its eye spots.

THE BEE DANCE

One of the most fascinating examples of communication is to be found among honey bees. Hive members returning from a foray tell other bees about the location of good sources of food, by performing a kind of dance. The speed and direction of the dance indicate the position of the food very accurately. If the food is within 100 metres or so the bee performs a round dance — quick circular motions made first in one direction and then the other. If the food is further away she performs a waggle dance — a figure-of-eight performed while waggling her abdomen. The angle at which she performs the dance tells other bees about the direction of the food.

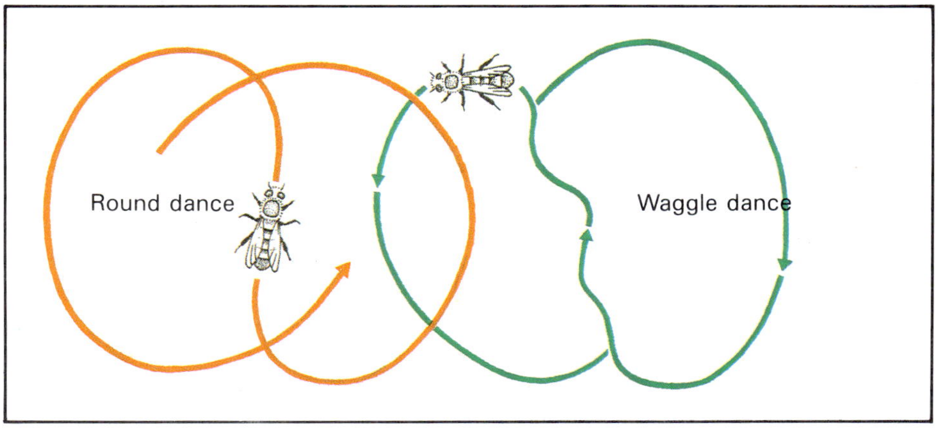

Right: Honey bees returning to their hive after a food foray inform other hive members about the location of food, by performing a sort of dance.

Attack and Defence

For many animals, life in the wild is harsh. Most grazing (plant-eating) animals are preyed upon by meat eaters. Even animals which eat other creatures may themselves become a meal for a bigger predator.

Animals which attack others in order to eat them use many different methods to catch their prey, and an array of weapons with which to overpower them. For example, large predatory mammals such as lions, tigers and cheetahs have powerful limbs, and jaws armed with sharp teeth. Most big cats chase their prey before knocking it to the ground with a blow from their paws. Then they bite the victim in the neck, quickly killing it.

Sometimes animals attack each other for reasons other than to catch food. During the mating season particularly, animals will fight, occasionally to the death, over a mate or a territory.

Above: The trapdoor spider lies cunningly concealed in a trap, waiting for prey. As soon as a likely meal passes, the spider rushes out and overpowers its victim by injecting it with a deadly venom.

Above: The peregrine (left) dives from the air on to its victim. It can fly at such great speed that few birds can escape it. The polar bear (right) catches seals which emerge from ice holes to obtain air.

Left: Dragonfly larvae have special jaws called a mask with which they grab their prey.

Left: Lions hunt their prey together. Some of the lions chase the prey to where other lions are lying in wait for them.

There are many ways in which animals try to avoid being eaten. Some animals are poisonous or distasteful to eat, and warn would-be predators of this fact by wearing bright 'warning colours'. Others are just too big to be eaten; no predator could attack and kill a fully grown rhinoceros or elephant.

Some creatures are quite able to look after themselves. Many types of antelope have sharp horns which can cause severe wounds to unwary predators. Others live in a herd, and are able to run fast. Quite often the predator will single out the weak or old members of the herd and leave the stronger members to escape.

Many creatures, particularly insects, rely on camouflage or a secretive way of life to avoid being spotted by predators.

Mountain hare (summer)

Mountain hare (winter)

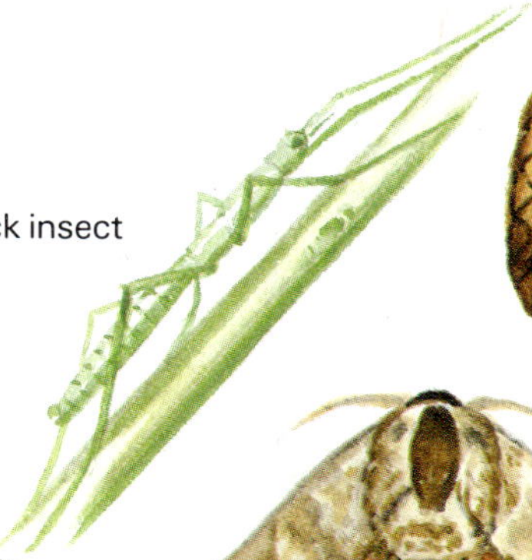

Stick insect

Pangolin

Eland

Porcupine

Hawkmoth

Left and above: Animals like the eland have sharp horns which can inflict a severe wound on predators. The porcupine is armed with a great number of spiny quills. These sharp spines give a painful lesson to any animal attempting to eat the porcupine.

Above: The mountain hare has protective colouration. In winter its coat turns white to match the snow. The stick insect relies on being mistaken for a twig. The pangolin curls into an armour-plated ball, while the hawkmoth uses the 'eyes' on its wings to alarm would-be predators.

Friends and Foes

Above: The speedy cheetah makes a dash to catch an antelope. But if the antelope can keep running the cheetah soon tires and gives up the chase.

Above: The pangolin, or scaly anteater, rolls itself into a ball when danger threatens.

Below: The Australian frilled lizard tries to frighten away its attackers by hissing and spreading out a fold of skin.

Animals in the wild have few friends, and many enemies. On land, the grazing animals, such as deer, antelopes, elephants and bison form large herds. A herd itself is a protection against danger, but often the stronger animals deliberately guard the weaker; for example, when musk oxen form their protective circle they place the younger animals in the centre.

In the sea, fish often stay together in huge shoals, and so do many species of whales. Whales and dolphins have been seen supporting and escorting an injured companion.

Animal Partners

Some of the most remarkable 'friendships' are those between animals of different species. They live together for mutual support, a state known as *symbiosis*. The hermit crab lives in a discarded shell, and has a sea anemone on the outside of the shell. The stinging cells of the sea anemone

protect the crab, and the anemone benefits from food particles discarded by the crab. When the crab grows and moves to a bigger shell, it takes its sea anemone with it.

Little fishes known as wrasses living in the Indian and Pacific oceans act as cleaners for bigger fish. They eat parasites, such as fish-lice, on the bodies of the larger fish, which queue up to be cleaned.

When Danger Threatens

Danger in the wild comes from the *carnivores*, the flesh-eaters, who prey on the *herbivores*, the plant-eaters. The term 'carnivore' includes not only such large animals as tigers and hyenas, but also many birds which prey on small creatures such as insects.

Some animals, such as the springbok, rely on their speed to escape from danger. The hedgehog curls up, presenting a predator with an unappetising ball of prickles. Many animals have *protective coloration*: their colouring matches that of their background. For example, a white Arctic hare is hard to see against snow. Small creatures often look like something else: there are moths which resemble bird droppings, insects which look like leaves or twigs, and plaice which match the sea bed on which they lie. Many creatures, such as wasps, are harmful to eat, and have *warning coloration*, generally red or yellow and black. Other animals mimic this coloration, even though they are not unpleasant to eat; for example, hoverflies look very like wasps.

Above: Some examples of insect camouflage on a tree trunk. At the top a katydid nestles among leaves of a similar colour. The stick insect poses to look just like a twig on the side of the tree, while the treble bar moth has colouring like the trunk itself. The leaf-insect 'hides' by sitting openly on a leaf. Even bright colours can be a camouflage, as can be seen in the yellow crab spider crouched on a buttercup awaiting its prey.

Right: Oxpeckers are African starlings which spend their lives living on large animals, such as the Cape buffalo shown here. They feed on ticks and other parasites which they pull out of the animals' skin. They warn their hosts when danger is about by uttering loud cries and flying about in the air. The large animals take no notice of their companions except when they give the alarm. The birds make their nests in holes in trees, or in rocks.

Animals and their Young

Above: Many species of snakes coil themselves around their eggs. This helps to protect the eggs from predators, but also keeps the eggs warm during incubation.

Animals must reproduce in order that each species can continue. The simplest animals, like the amoeba, reproduce by simply splitting into two. Most animals, however, reproduce by means of a male and female mating together.

For many animals, the first task in reproduction is to find a safe place to lay their eggs or give birth to their young. In the insect world, many species lay their eggs in holes or crevices, or glued safely to the underside of leaves away from the eyes of predators. Some insects, for example some kinds of wasps, even put in a supply of food ready for the young to eat when they hatch.

Other species of animals may have to fend for themselves as soon as they are born.

Below: Mammals, like the opossum shown here, feed their young with milk, produced in the mammary glands of the female.

Above: Many insects undergo a four-stage life cycle. The adult lays eggs which hatch into larvae. The larvae feed and grow, and turn into pupae. From the pupae emerge new adults.

Right: A reed bunting feeding its chicks at the nest.

Many kinds of birds build nests in which to lay their eggs and rear their young. During spring, both parents will often work ceaselessly hour after hour, building an elaborate nest of twigs, grass, leaves and down so that their chicks will hatch safe from enemies such as snakes or other birds.

Some animals display a remarkable degree of parental care towards their young. We often think of crocodiles as ferocious animals but, as soon as the young hatch from their eggs, the mother crocodile carries her babies gently in her mouth to a safe place.

Once born, the job of finding food begins. Although many creatures must find food for themselves when they are born, most of the more advanced animals, such as birds and mammals, rely on their parents to provide nourishment for them until they have grown big enough to feed themselves.

Above: Scorpions show a remarkable form of parental care. When the young have hatched, they climb on to their parents' back and are carried about, safe from danger.

Below: Many animals, like these brown bear cubs, are taught how to hunt by their parents. Play fighting also helps them to fend for themselves in the wild. The mother watches over the cubs while they play.

Animals on the Move

Each year many animals make long journeys in search of food, shelter from harsh weather or a place to breed. These journeys are called migrations. Among the best-known migrations are the annual journeys from Europe to Africa made by cuckoos, swallows, warblers and many other birds.

Many animals live in parts of the world where the climate changes throughout the year. In the temperate zones these changes are regular and are called seasons, although in the tropics there is little change from one month to the next. As the seasons change the conditions can become less suitable for feeding or breeding; some animals can cope with these changes,perhaps by hibernating, but others overcome the threat of harsh conditions by moving to more favourable areas.

Insect Migration

We normally think of migration as being undertaken only by larger animals, but many insects, especially some butterflies, undergo migrations as lengthy as those carried out by birds and mammals.

The red admiral is a colourful butterfly which is found in north Africa and the Mediterranean. Each spring, however, some of them move northwards into Europe and breed as far north as England and Scandinavia. The following autumn, a few of the new generation fly south to escape the onset of the cold weather. Although many perish on the way, some survive the return journey — an impressive feat for an insect.

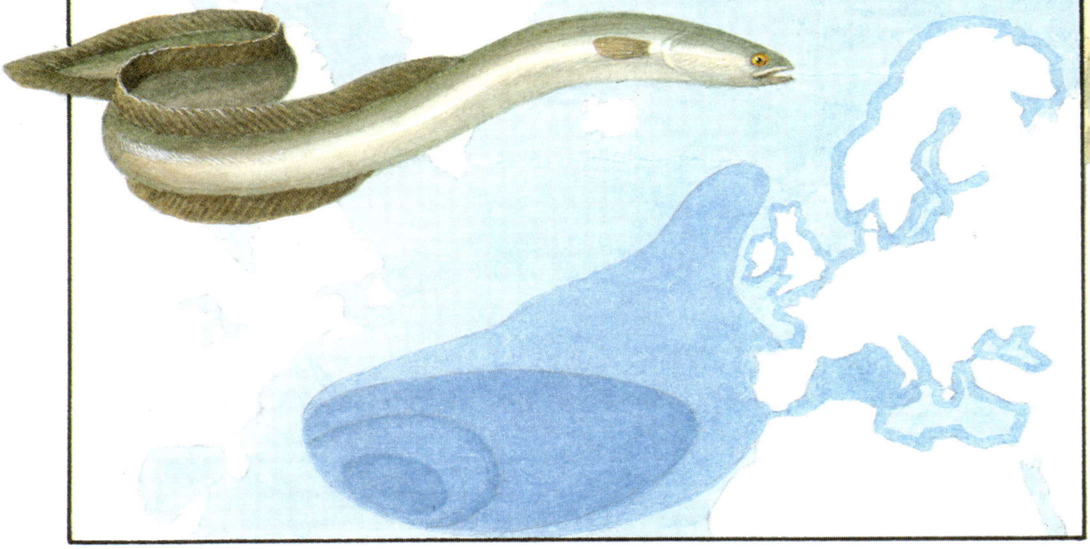

Above: The herds of wildebeest on the African plains follow the rains. In this way they can eat the new growth of grass. The animals are constantly on the move, covering hundreds of kilometres a year.

Left: European eels are spawned in the Sargasso Sea. The Gulf Stream carries the young eels to our shores and they move into fresh water. After about ten years they return to the sea to breed.

Above: In some years lemming numbers reach plague proportions. To escape the overcrowding many go on mass migration. At this time predators such as snowy owls and foxes benefit from their abundance.

Below: Bird migrations often involve vast distances and require spectacular navigational skills. Many migrate from northern regions to more temperate areas. Some, like arctic terns, cross to the southern hemisphere for the winter.

Migration of Larger Animals

Both marine mammals as well as land-living species undergo migrations. Seals leave their breeding grounds in spring, and return to warmer seas to spend the winter. When feeding becomes poor many mammals just keep moving until they find better conditions. Herds of animals in Africa follow the rains, grazing the fresh growth of grass that follows. Some reptiles also migrate. Marine turtles haul themselves on to the same beach every year to lay their eggs.

The movements of birds have evolved to become extremely predictable, and they often follow distinct routes over long distances. The arctic tern is famous for its travels from its breeding grounds in the arctic to its wintering area in the antarctic. In its travels it covers over 32,000 kilometres a year, and probably over 160,000 kilometres in its lifetime. As well as travelling further in its life than most other living creatures it also sees more daylight than species which remain in one area.

Nocturnal Life

When darkness falls over the town and countryside, another world comes to life — the world of nocturnal plants and animals. The animals which had been active during the day seek safe places to shelter for the night, and other creatures take their place.

When the sun goes down the temperature drops, and moisture forms on the ground and in the air. These are ideal conditions for the small creatures which thrive in a damp environment. Earthworms, insects, spiders, centipedes and other small invertebrates creep from their hiding places to find food or to mate.

In the darkness other, larger creatures feel safe, too, and leave their burrows to hunt. Mice and voles search for insects, seeds and other food items. They in turn fall prey to the hunters of the night — the predatory animals such as foxes, stoats and owls. Larger creatures such as deer leave the safety of the woodland to graze in the clearings.

Most nocturnal animals are specially adapted for their night-time existence. Badgers and deer have a well-developed sense of smell, for instance. Deer also have acute hearing. Moths can detect the scent given off by females from great distances. Owls swoop down on their prey without warning, for their wings make scarcely a sound as they beat.

Animal Architects

Above: The Great Barrier Reef is the only structure built by animals which can be seen from space. It was constructed by tiny marine organisms called coral polyps, like the ones seen here, which encase themselves in a tough chalky shell.

When we visit a cathedral we marvel at the skill of the men who constructed it. But if we consider the building feats achieved by some animals, often working entirely alone, some of these are equally amazing.

Animals build for similar reasons to people; to make a shelter from the elements, protection for themselves or their young, or in order to catch food. Animal shelters and homes range from the little tubes of caddis larvae, to the often complex nests which birds build for their eggs and young.

Some animal constructions are quite small, and are designed to accommodate only their makers. Others, such as termite mounds, are enormous, and provide a home for thousands of individuals. Building material also varies. Some animals use materials they gather from their surroundings, and others produce the building material themselves. Spiders, for instance, make their webs from the silk which they spin inside their own bodies.

Nests

Birds are among the most familiar of animal architects. At the start of each breeding season, many species of birds build a nest in which to lay their eggs and rear their young. Many nests are quite simple structures consisting of interwoven twigs or grass stems. Sometimes these are neatly arranged, and in other species they are a jumbled mass of material. Often these nests are placed in the fork of a tree or in a crevice, giving the nest support.

Nests are usually built from material which the birds gather close by. This makes the nest quick to build, as well as helping it blend into the surroundings. The long-tailed tit's nest is particularly difficult to spot, since the birds use a mixture of lichens, spider's silk and dry grass.

Below: Many spiders produce sticky silk which they spin into beautiful but deadly webs. These trap flying insects which the spider quickly bites and paralyses. The victim is then wrapped in silk, ready for eating later.

Other Homes

The larvae of the caddis fly are also skilled architects. These soft-bodied freshwater insects protect themselves by building tough tubes in which to live. Each species has its own favourite building material; some use plant leaves, others use sand or snail shells. The tube is open at one end, and the larvae drag themselves around using their front legs.

Termites are tropical insects which live in large colonies. Their mounds, which are usually made from soil, are very strong. They can also become very large, often exceeding 6 metres in height. This feat is all the more remarkable when you consider that the animals which built them were less than 1 centimetre long.

Above: Weavers are birds which build nests together in thorn bushes. The nests are very carefully woven from grass, and often have long entrances designed to prevent predators getting in.

Above: Many wasps are social insects which build papery nests in which to rear their young. They construct layers of hexagonal-shaped chambers in which the eggs are laid and the young wasp larvae grow.

Below: The beaver lodge is a remarkable structure of twigs and logs with its own underwater entrance. Built on a river, the beaver constructs a dam to raise the water level around the lodge.

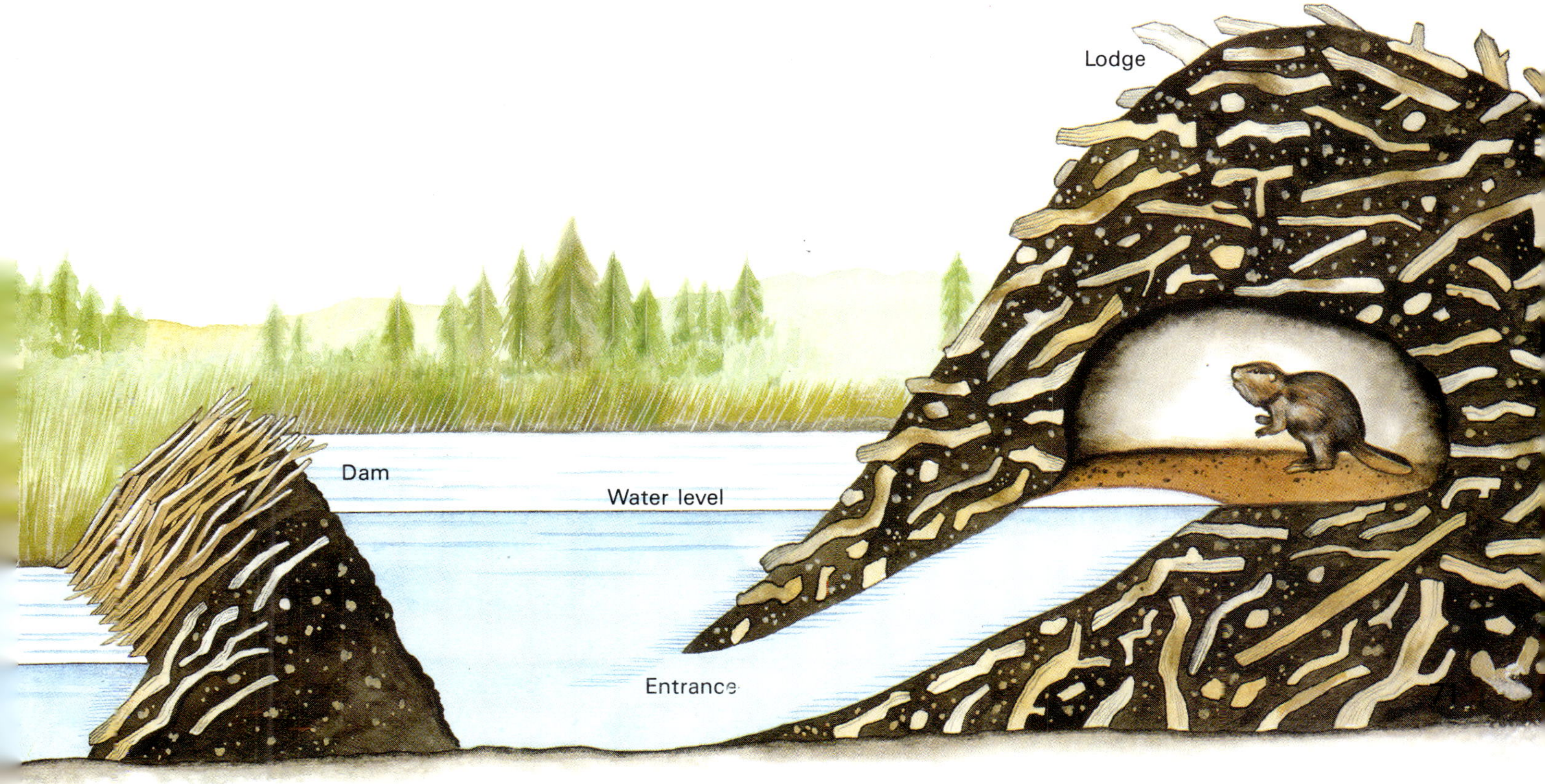

Living Together

We have already seen how the world of plants and the world of animals are closely interwoven together. Without plants, no animals could survive. Many animals would have no source of food or shelter, and this would mean no food for the other animals which prey upon them. Likewise, many plants rely on animals for their survival. Without insects and other animals, many flowers would not be pollinated nor would their seeds be dispersed. Insectivorous plants such as the sundew and the Venus fly trap must also catch and digest insects and other small creatures in order to obtain the nitrogen which is essential for their healthy growth.

Many animals also need the help of other animals to help them avoid predators or to help them find food. For this reason some species join together in great herds. Sometimes, however, animals which are quite unrelated join together to improve their chances of survival.

Below: Grazing animals, like these herds of buffalo on the American plains, join together in great herds for protection. The herd guards the youngest members which might fall victim to predators, and each animal is always on the lookout for any signs of danger.

PLANT AND ANIMAL PARTNERSHIPS

Although most of the partnerships in nature are between different species of animals, or between different species of plants, there are some partnerships in which a plant and an animal live together. Many aquatic animals, such as corals or the freshwater creatures known as Hydra, contain green algae within their bodies. The algae have a safe place in which to live, and in return they provide oxygen and absorb the various waste materials produced by the animals. Within the bodies of many animals live bacteria. These help to break down the cellulose found in plants and so enable the animals to digest grass.

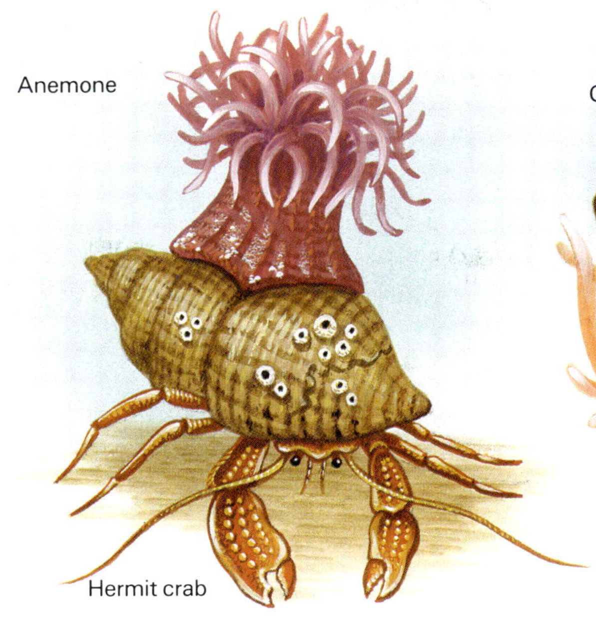

Above: The hermit crab often carries an anemone around on its shell. The anemone protects the crab, and in return shares some of the crab's food.

Above: Clown fishes are brightly coloured inhabitants of coral reefs. They feel quite safe swimming among the deadly stinging tentacles of anemones which also inhabit the reef. The anemones let the clown fishes swim among their tentacles to escape predators, in return for ridding them of parasites.

Animal Partnerships

In the Animal Kingdom there are many examples of animals living together for the benefit of one, or both, partners. The simplest form of this behaviour is when one animal spends its life living close to another animal for its own benefit. Birds such as cattle egrets ride on the backs of grazing animals as they walk slowly through the long grass. As they walk, they disturb insects and other creatures which the egrets then swoop down to feed on. Although the grass-eating animals do not themselves benefit from the association, they do not suffer either.

Sharks are often attended by fishes known as remoras. The remoras attach themselves to the sharks by means of suckers on their heads. They eat the scraps left over by the shark when it feeds, which in return uses the remoras as a 'vacuum cleaner' service, for they remove parasites from the shark's skin.

Sometimes two different species of animals rely so much on each other for food or protection that they are unable to live so successfully on their own. This condition is know as symbiosis. There are many examples of symbiosis in the Animal Kingdom. The partnerships shown on this page between the hermit crab and the anemone, and between the clown fishes and the anemone are two.

Top left: Oxpeckers are African birds which are allowed to feed on the skin parasites of rhinoceroses. In return, the oxpeckers alert the rhinoceros of danger.

Left: Spur-wing plovers and dikkops help keep crocodiles free of parasites.

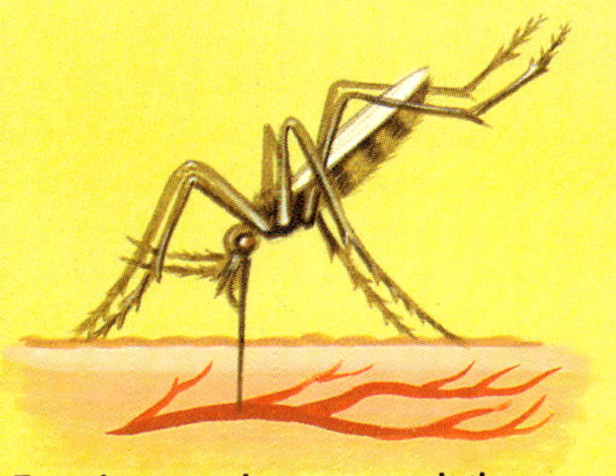

Female mosquitoes transmit the disease called malaria. The mosquitoes feed on the blood of man, which they obtain by inserting their mouthparts into blood vessels near the skin. As they do so, they inject into the bloodstream tiny, single-celled parasites called trypanosomes. Trypanosomes enter the red blood cells of humans and multiply. After a time, they burst from the cells and into the bloodstream, producing the familiar malarial fever. Some of the trypanosomes pass to another mosquito when it bites a victim, ready to be passed on to another human.

Animal Curiosities

Throughout the Animal Kingdom there is an incredible array of sizes, shapes, colours and behaviour. At first glance some of these animals may seem particularly strange, but in nature there is always a good reason for any peculiarity. What, to our eyes, may seem strange, is in fact essential for an animal's survival in its chosen environment. The more we study nature the more we realize that there is no 'ordinary' animal. Each species has evolved its own shape, colour and behaviour to ensure its survival.

Nevertheless, there are some animals whose appearance and life styles almost defy the imagination. There are fishes that fly and others that walk on dry land, mammals that can fly in complete darkness, spiders that spin webs under water and many more extraordinary creatures.

Below: Many animals hide from predators, but not so the colourful sea slug which actually advertises its presence. The bright colours warn predators of its unpleasant taste.

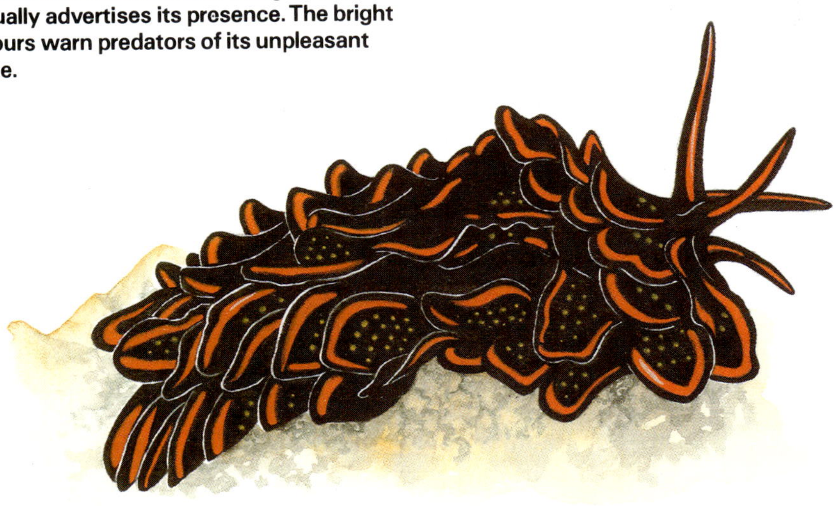

Unusual Spiders

Many spiders spin silk webs to catch their prey. As if this extraordinary feat were not enough, some use their silk in even more unusual ways. For example, the spitting spider does as its name suggests; it spits silk over its prey until it is so entangled that it cannot escape.

The purse web spider does not spin an ordinary web at all. Instead, it lives under ground in a silken tube strengthened with plant fibres. Part of the tube lies on the ground, however, and when a passing insect crawls over its surface, the spider rushes up the tube and grabs the victim through the tube. Needless to say, the tube needs considerable repair after this!

Below left: The water spider spends its life under water, constructing air-bells in which to live and lay its eggs.

Below right: The crab spider pretends to be the inside of a flower. When an insect lands, it is grabbed by the waiting spider.

Odd-shaped Eggs

Bird's eggs are sometimes brightly coloured and sometimes camouflaged to escape the attention of predators, but nearly all eggs are the same shape. The egg of the guillemot is different, however. Its egg is pear-shaped, being blunt at one end and pointed at the other. Guillemots breed on sea cliffs and lay their eggs on steep ledges. The peculiar shape of the egg prevents it rolling off the edge. If it becomes dislodged it just rolls round in a circle.

Above: Flying fish can glide for over 100 metres, using their specially developed fins. This ability helps them escape when being pursued by predators under the water.

Below: Until 1938 the coelacanth was thought to have been extinct for 60 million years. This 'living fossil' is now known to live in deep water in the Indian Ocean.

Coelacanth

Below: The goliath beetle of Africa is the heaviest insect on earth. Weighing over 100 grams it is heavier than many small mammals and birds.

Goliath beetle

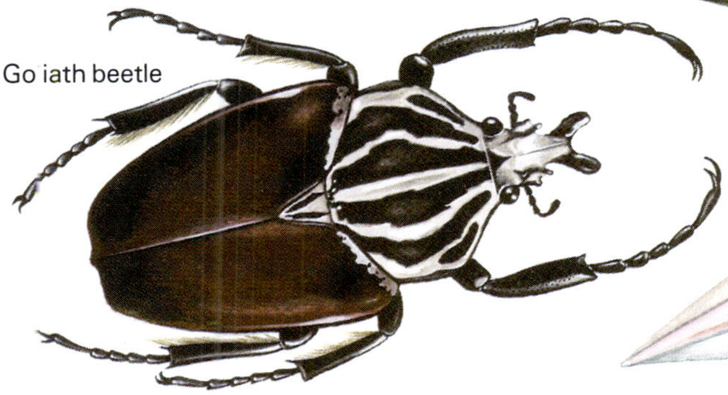

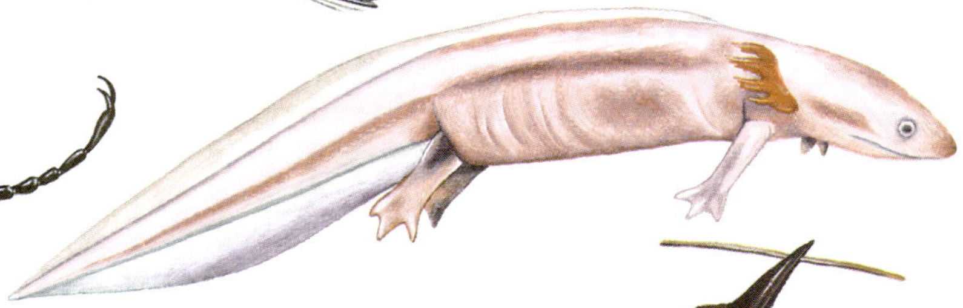

Axolotl

Above: The axolotl is a remarkable amphibian related to salamanders. It lives its life as a permanent larva, and uses feathery gills with which to breathe. Even more remarkably, it can even breed while still a larva.

Below: Most species of bat are able to fly in complete darkness, easily avoiding obstacles and catching insects on the wing. They use echolocation to find their way around, although they are not in fact blind.

Cactus wren

Above: The cactus wren is a rare example of an animal which uses a tool. By holding a stick in its beak, it prises insects out of crevices in bark.

The Plant Kingdom

There are nearly half a million species of plants on the earth. With the exception of some primitive types, they all need three basic ingredients to survive: air, light and water. Plants are found in almost every type of habitat, including most of the surface waters of the world's rivers, lakes and oceans. Most species live in temperate and tropical regions but plants are extremely adaptable, and have even conquered such inhospitable environments as hot mineral springs and dry deserts with little rainfall.

How Plants Live

Although there is a great variety in the shape, size and appearance of plants, all of them, apart from fungi and bacteria, make their own food using simple raw materials and energy from the sun. They trap the sun's rays using the green pigment chlorophyll, and combine water and carbon dioxide (a gas present in the air) to make simple sugars. They also release oxygen into the air — which all living things need for respiration. This process, known as photosynthesis, is vital to the survival of all life on earth.

Plants include some of the smallest things as well as the largest. The surface waters of ponds and lakes teem with microscopic algae, many of which exist as just a single cell. By contrast, the giant redwood trees of California may grow over 90 metres tall.

Some plants reproduce so rapidly that new individuals are produced within minutes by simple division, whereas the bristlecone pine trees live to be more than 3,500 years old. Each year they still produce a new crop of seedlings.

Almost all types of plants are eaten by one species of animal or another. Animals that eat plants are called herbivores, and they include such creatures as cows and locusts. Other animals eat only meat and these are called carnivores. However, the animals that they prey upon will have eaten plants. If it were not for plants, animal life could not exist on earth.

Right: This evolutionary 'tree' shows how the different groups of plants have evolved throughout the geological ages.

Dicotyledons
Ginkgo
Conifers
Tertiary/Quaternary Period
Monocotyledons
Cretaceous Period
Cycads
Jurassic Period
Ferns
Angiosperms
Gymnosperms
Triassic Period
Horsetails
Permian Period
Mosses
Liverworts
Club mosses
Carboniferous Period
Brown Algae
Fungi
Lichens
Devonian Period
Bryophytes
Fungi
Green Algae
Silurian Period
Red Algae
Fungi
Ordovician Period
Golden Algae
Slime moulds
Bacteria
Cambrian Period
Blue-green Algae

Lowly Plants

The lowly plants have existed on earth for much longer than the flowering plants we grow in our gardens. Many have remained almost unchanged for millions of years. They generally have a simple structure and, with the exception of ferns, do not have supporting fibres. This means that they cannot grow to any great size.

Most of the lowly plants contain the green pigment chlorophyll. This helps them to trap sunlight energy and make their own food. The exceptions are the bacteria and fungi which have to use other sources of food such as the dead bodies of plants or animals, and are responsible for decay.

Advanced plants mostly have flowers and produce seeds in order to reproduce. Lowly plants, however, often reproduce by means of spores. The bacteria and many of the algae reproduce by simple division, and like the rest of the lowly plants they need damp conditions.

Left: Mushrooms and toadstools are the reproductive parts of fungi. Most of the fungus lives underground forming a complex web of thread-like strands called a mycelium.

The spores of fungi are always present in the air. They grow and thrive when food, such as bread, is left in damp conditions, forming patches of mould.

Algae

Algae are the simplest forms of true plant life. They possess chlorophyll, but many contain other pigments, giving them many colours. They range from simple, single-celled plants which make ponds appear green, to immense seaweeds, several metres long.

Fungi

Unlike algae, fungi cannot make their own food and depend on a ready-made source. Many feed on dead matter and cause decay.

Lichens are strange plants which are the result of a partnership between algae and fungi. The relationship is called 'symbiotic', and both partners benefit from it. Their growth is very slow.

Mosses and Liverworts

Mosses and liverworts are simple plants which lack true roots and stems. They are restricted to damp habitats and reproduce by spores produced in capsules.

Ferns

Ferns are generally long-lived plants. They have a rootstock and a strong, supporting stem which can conduct water and dissolved food substances. This means that they can often reach a large size. They still produce spores during reproduction and depend on water for part of their life cycle.

Below: Fern spores are spread by the wind

Above: Ferns come in a variety of forms. Some are rooted in the soil and grow to large sizes. Others colonize walls or even grow on other plants, but all need damp conditions.

Right: Seaweeds are algae which grow on rocky shores. Mosses are common on walls and in damp places.

Flowers

Rose

Primrose

Flowers contain the male and female cells of a plant. For successful fertilization to take place, the pollen (male cells) from one plant has to reach the ovum (female cell) of another. However, plants have one basic problem—they are rooted to the ground. To overcome this, plants have evolved all sorts of ways of transferring the pollen.

Despite their unlikely appearance, grasses are true flowering plants which produce countless thousands of minute pollen grains that are carried by the wind. Not surprisingly, most end up in the wrong place but enough reach another grass flower to ensure their survival. More typical flowers have colourful petals and often a strong smell to attract insects. In return for a meal of nectar, the insects transfer the pollen to the next flowers which they visit.

Flower Plan

Although some species of flowers sometimes contain only the male or the female cells, most contain both but go to great lengths to prevent self-fertilization.

Despite a great variety of appearances most flowers conform to a basic plan with sepals, colourful petals, stamens which produce the male cells called pollen, and a central stigma containing the female cell or ovule.

The parts of a flower

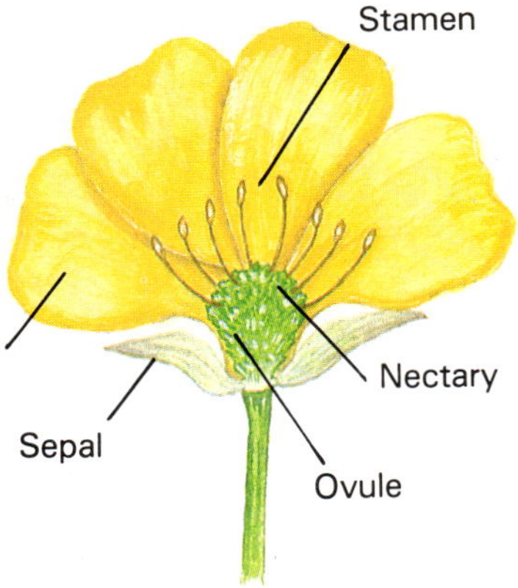

How a plant disperses its seeds

Above: Coconut seeds float in the sea.

Above: Willowherb seeds are carried by the wind.

Above: Burdock seeds have hooks which catch in the fur of animals.

Above: Poppy seeds are shaken from the pod.

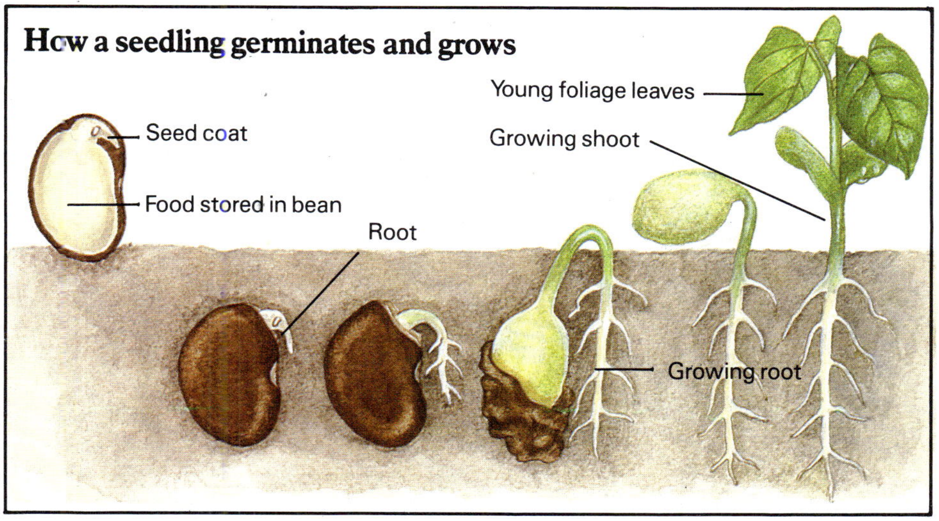

Above: Flowers occur in a whole variety of shapes, sizes and colours. They range from the inconspicuous flowers of grasses to the large and showy flowers of the sunflower. Some plants, such as the cactus, flower only briefly and irregularly, whereas others are in bloom for a full season—or even continuously.

Trees

Trees, woods and forests cover nearly one quarter of the earth's surface. They are a very important part of our environment, for they provide food, homes and shelter for a great variety of other living creatures. The key feature of all trees is their central woody stem or trunk. This grows as the tree grows, and provides support and protection.

Trees can broadly be split into those which drop their leaves each year (called deciduous trees) and those which do not (called evergreens). Many evergreens bear their seeds in cones and have needle-like leaves; these trees are called conifers. There are also some broadleaved trees, such as holly, which do not shed their leaves each autumn.

Palms are a special group of trees. Although they are deciduous, they are placed in a group of plants called monocotyledons, the group which also includes the grasses.

Above: The hazel is a deciduous tree. It has tiny female flowers, and male catkins which produce pollen. The seeds are encased in a tough shell, or nut.

Right: By counting the rings on a cross-section through a trunk you can estimate the age of a tree. Each ring roughly corresponds to a year's growth.

Below: There are many different types of tree throughout the world. Here are some typical species.

How Trees Grow

As well as providing support for the tree, the trunk carries the water supply for the leaves of the tree. Although the centre of the trunk is usually tough and woody, the outer tissues consist of tubes through which water is transported from the roots to the leaves. Some of this tissue also transports food from the leaves down to the roots and other parts of the tree. The trunk tissues are protected by the bark. If this is damaged, the tree may die.

Leaves are the vital life-supply for all plants. They produce food by photosynthesis, using the energy of sunlight to combine water and carbon dioxide. There are many different shapes of leaves, but all are designed to make best use of the available light.

In some parts of the world, the climate does not change during the year and the tree grows continuously. But elsewhere the climate varies, with either cold and hot seasons or dry and wet periods. The tree grows best in favourable conditions and, as a result, we can see growth rings in the trunk, each dark ring corresponding to a year's growth. In a good year the ring is thicker than in a poor year. By counting the number of rings, the age of the tree can be worked out.

In order that trees may reproduce, the female part of the flower must first be pollinated. In some trees, male and female parts are in the same flower. In others, they are in separate flowers or, sometimes, on separate male and female trees, such as holly.

Above: The spruce is a conifer with tiny flowers. The fertilized seeds grow and ripen in cones.

Right: An oak tree supports a whole community of plants and animals. Some animals feed on its leaves or acorns while others in turn prey upon them. Many beetles and fungi feed on the decaying wood and leaves.

Records in Nature

HIGHEST MOUNTAINS (metres)

Above: During the Earth's long history, many mountain ranges have been created by movements in the rocks, and then gradually worn away by water, wind and ice. The highest peaks today are in the youngest mountain ranges – the Himalayas, Andes, Rockies and Alps. There may well have been higher mountain peaks millions of years ago.

There is a great range in the size of living things. At one end of the scale there are viruses – organisms on the very borders of life – which are too small to be seen without the aid of powerful electron microscopes. At the other extreme there is the blue whale which can weigh over 120 tonnes, and the Californian redwood tree which grows as high as a 35-storey building.

The life-span of living things also varies greatly. Small insects, such as the mayfly, have an adult life-span which can be measured in days or even hours, while the tortoise may live for 150 years or more. Some plants, especially trees, live far longer. There are giant sequoia trees in California which began life almost 4,000 years ago, and bristlecone pines which are as old as the pyramids.

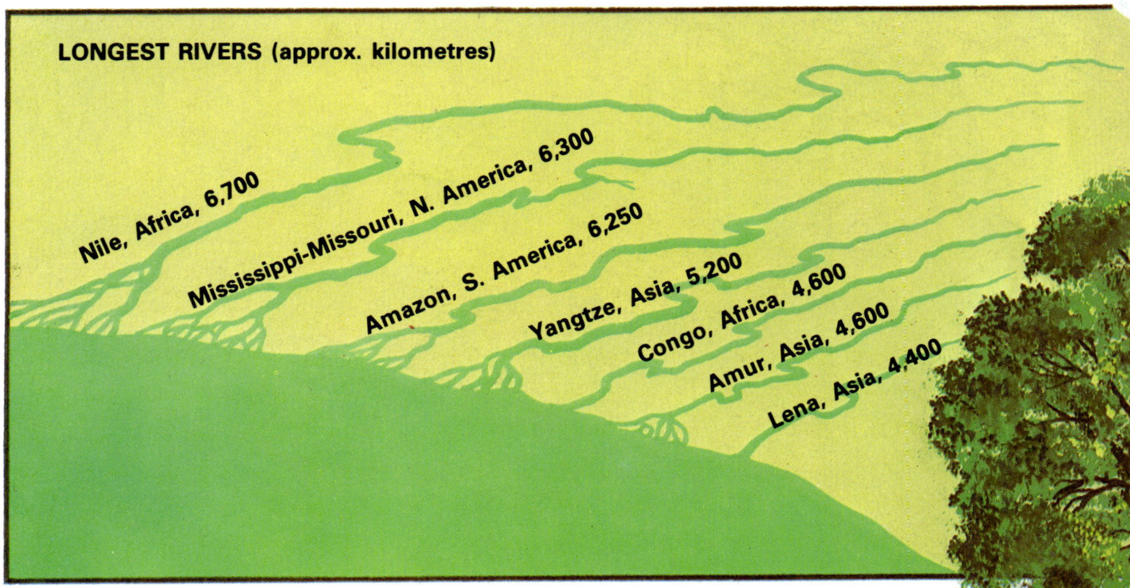

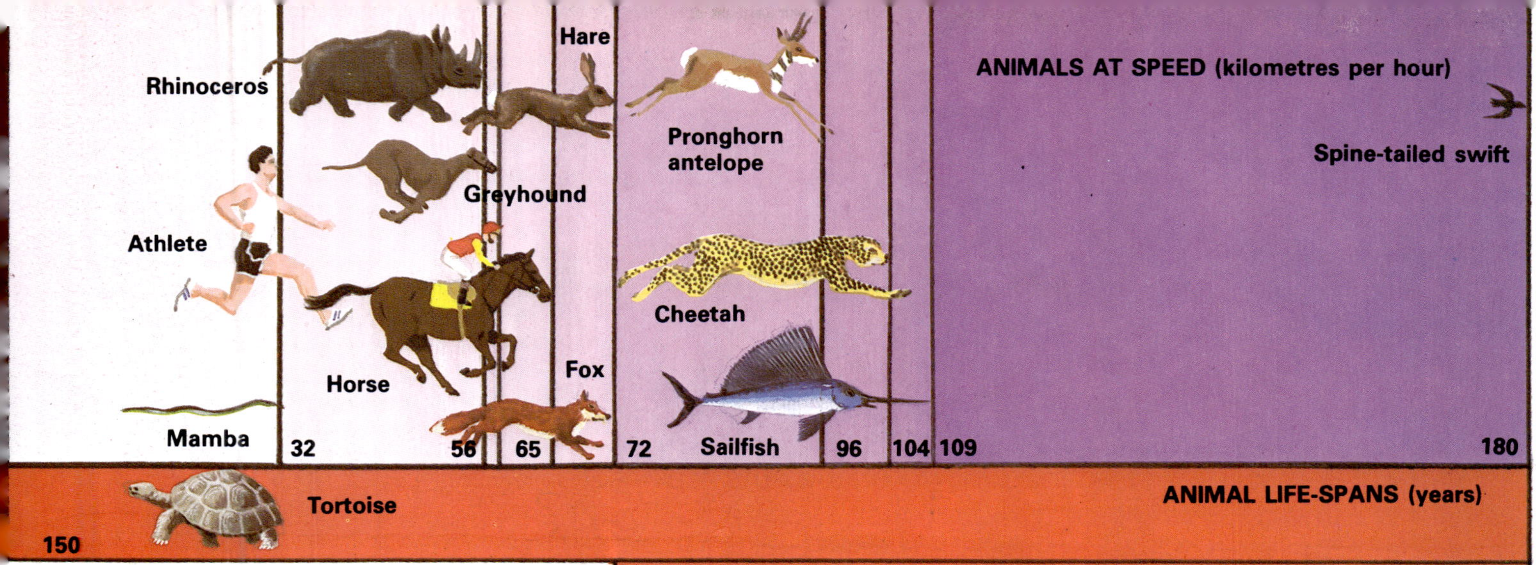

The table shows the average maximum life-spans of various animals. It is based upon records of animals in captivity. Animals in the wild have shorter life-spans.

The speed at which animals move is often difficult to measure accurately, but there is no doubt about the record holder. The spine-tailed swift can fly at more than 180 km/h. The fastest land animal is the cheetah, which can exceed 100 km/h over short distances. The proverbial tortoise does well to reach 1 km/h.

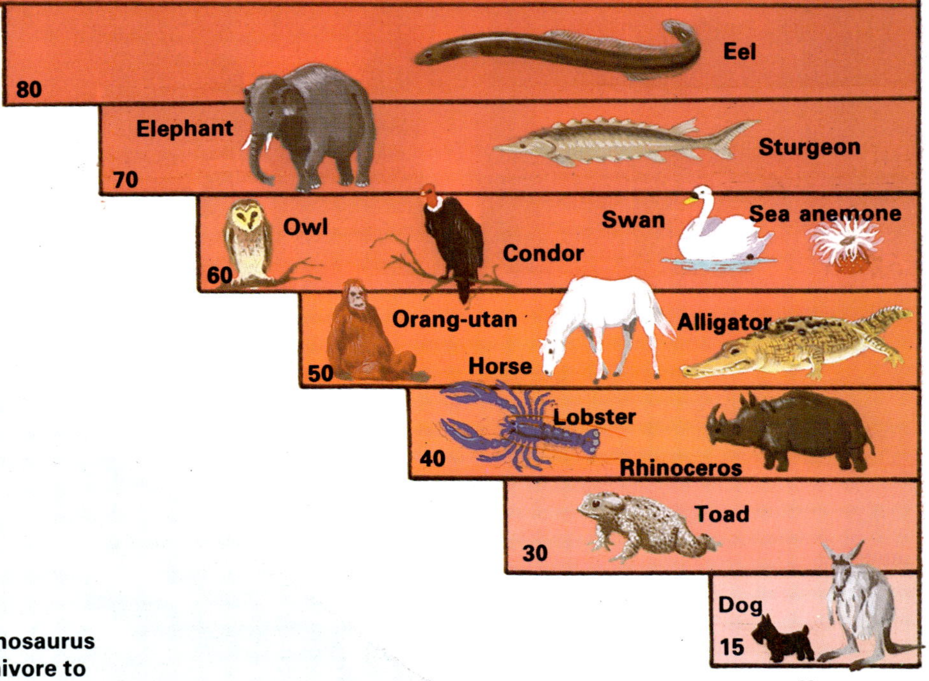

Some mayflies have an adult life of only a few hours.

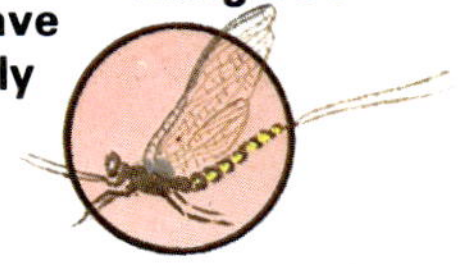

Mayfly

The dinosaur Tyrannosaurus was the largest carnivore to roam the Earth. It weighed more than a present day elephant and had teeth over 15 cm long.

Blue whale

The elephant is the largest land animal. It can weigh up to 11 tonnes.

The giraffe is the tallest animal, reaching 6 metres.

The ostrich is the tallest bird with a height of more than 2.5 m.

Wonders of the Plant World

In its own way, the Plant Kingdom is just as unusual and fascinating as the Animal Kingdom. Also, like the Animal Kingdom, the Plant Kingdom has its share of curious-shaped species, and plenty of species with strange and bizarre ways of life.

We have only to think of fungi, those peculiar plants which can grow without the aid of sunlight and seem to pop out of the ground overnight, to realize that some plants are very odd. However, as we shall see, there are others which can confuse scientists even more!

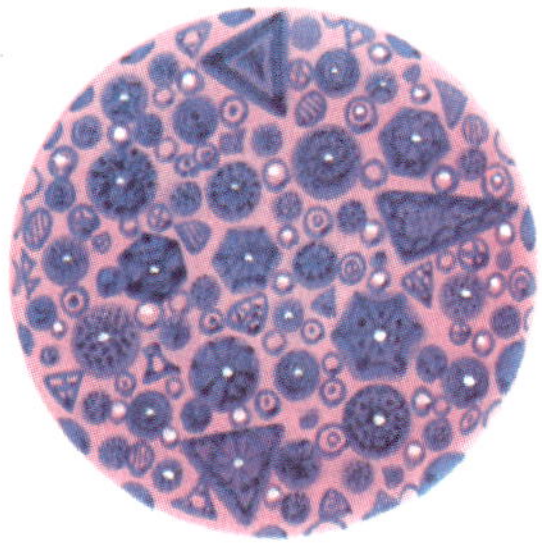

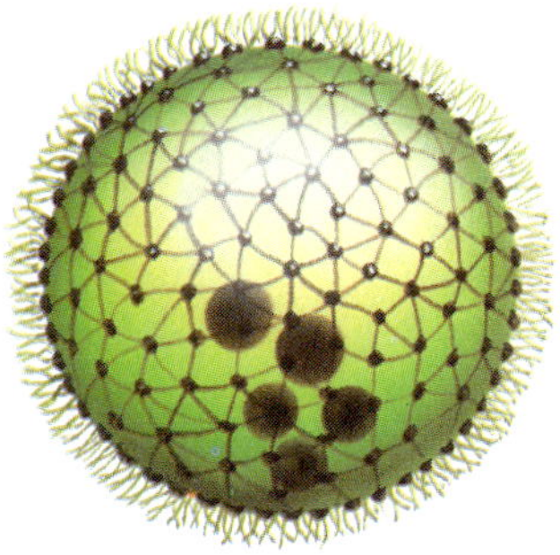

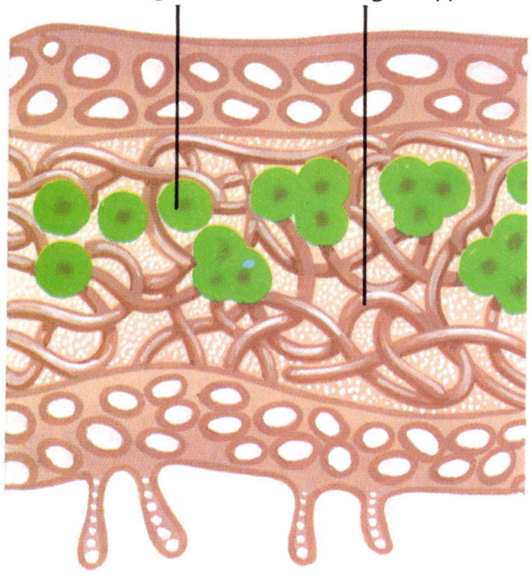

Left: Diatoms and colonies of single-celled algae like Volvox can move freely about in the water.

Right: This cross section through a plant stem shows how its water and minerals are absorbed by the suckers of mistletoe. Mistletoe grows on trees such as poplars.

Bottom left: Lichens are unusual plants made up of single-celled algae sandwiched between a fungus.

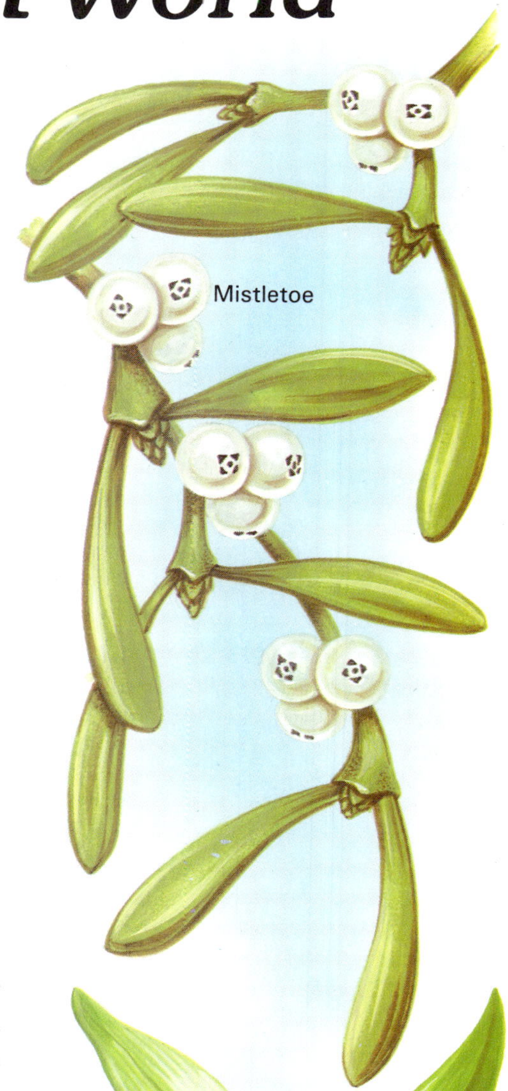

Lowly Plants

One of the ways that we normally use to tell animals from plants is by the fact that animals can move about, while plants stay fixed in one place. But among the group of primitive plants called the algae there are several species that can actually move about on their own.

Diatoms are tiny algae living within a silica shell, which can glide about in the water. Volvox is a plant which consists of a group of single-celled algae each of which has tiny hairs. By beating these Volvox can also move about in the water.

Flowering Plants

Among the flowering plants there are also many unusual species. Although most flowering plants make their own food by using the energy of sunlight trapped in their leaves, there are some kinds which 'steal' the minerals and food substances from other plants.

Dodder is a twining, parasitic plant whose stem sends suckers into the host plant.

Mistletoe is known as a partial parasite. It can make its own food using sunlight, but first it must rob a host plant of some vital minerals.

The most unusual flowering plants are the carnivorous species. They live in places, such as bogs, which lack the nitrogen they need for healthy growth. Therefore they catch insects and other small creatures in special traps (which are really special leaves) and absorb the nitrogen from their bodies.

Rafflesia

Giant redwood

Above: Rafflesia is the world's biggest flower. It can reach 60 centimetres across. Giant redwood trees can reach heights of 110 metres.

Venus fly trap

Carnivorous plants like the Venus fly trap and the pitcher plant eat small insects and other creatures.

Pitcher plant

Glossary

Abdomen One of the main parts of the body of an arthropod such as an insect. The abdomen is made up of the end segments of the body. In vertebrates, it is the part of the body containing the intestines.

Algae A large group of non-flowering plants. The biggest and most important algae are the seaweeds, but many single-celled species of algae also exist.

Amphibian A cold-blooded vertebrate animal descended from fishes. Most amphibians must return to water to lay their eggs. Common amphibians include frogs, toads, newts and salamanders.

Antenna One of the projections found on the head of insects and crustaceans used for sensing the environment. Some crustaceans also use their antennae for swimming.

Bird A warm-blooded vertebrate animal whose body is covered with feathers. Birds descended from reptiles and lay shelled eggs.

Bivalve A kind of mollusc whose shell is made of two parts. Common bivalves include mussels, cockles and scallops.

Breeding season The times of the year when certain animals come together for the purpose of mating and producing young.

Burrow The underground home of an animal.

Camouflage The ability of a living creature to disguise itself to avoid being seen by its enemies.

Carnivore A meat-eating animal, such as a lion or tiger.

Caste One of the different kinds of individuals in a social insect colony, such as a worker or a guard.

Caterpillar The larva of a moth or butterfly. The job of the caterpillar is to eat and grow, in readiness for turning into an adult.

Catkin A group of tiny flowers, usually hanging down from a main branch or stem.

Chlorophyll The green pigment found in all groups of plants except for the fungi. Chlorophyll is contained within cell structures called chloroplasts. It is used to trap the energy from sunlight to enable plants to make their own food.

Cone The reproductive structure of a conifer, bearing either female parts (ovules) or male parts (pollen). In dry weather the cones open to shed their pollen and enter the female ovules.

Conifer A cone-bearing tree, such as a pine, spruce or larch.

Coral A small, marine animal related to the anemones which builds around itself a chalky skeleton. Some corals live by themselves, but many form huge colonies.

Deciduous tree A tree which sheds all its leaves at a certain time of the year.

Dinosaur A member of a group of prehistoric reptiles which lived during the Mesozoic Era. Most dinosaurs were extremely large.

Dormant In a resting state. Many plants undergo a dormant period when conditions are unsuitable for further growth.

Embryo A stage in the development of an animal or plant following fertilization of an egg or ovum.

Exoskeleton The hard, protective outer covering of the arthropods. The exoskeleton is composed of a material called chitin.

Fang A long front tooth of a snake, or of a carnivore such as a lion.

Feather The special structures covering the bodies of birds. Feathers are mainly used for flight, insulation, camouflage and display.

Fish A cold-blooded, aquatic vertebrate animal whose body is covered with scales.

Flower The reproductive structure of flowering plants. Some species have flowers which contain both the male parts (pollen) and female parts (ovules), and others have the male and female parts on separate flowers.

Fossil The preserved remains of a dead animal or plant.

Frond The leaf of a fern, or the strap-like parts of a seaweed or lichen.

Germination The process by which a seed produces a root and shoot, eventually forming a new baby plant.

Gill A structure found in aquatic animals used for extracting oxygen from the water when breathing.

Habitat The particular place where plants or animals live, such as a seashore or a woodland.

Herbivore A plant-eating animal, such as a cow or zebra.

Holdfast The root-like structure by which seaweeds attach themselves to rocks.

Incubation The process of maintaining eggs at the correct temperature whilst they undergo development prior to hatching.

Invertebrate One of the large group of mainly small animals which do not possess backbones.

Larva The stage in the life cycle of some animals before they turn into adults. Larvae are usually capable of feeding for themselves, but look different from the adult form.

Lichen A slow-growing, non-flowering plant consisting partly of an alga and partly of a fungus.

Mammal A warm-blooded vertebrate animal which suckles its young. Common mammals include elephants, dogs, cats, monkeys and tigers.

Mammary gland Gland found in female mammals which produces milk for feeding the developing young.

Mesozoic Era A time in the earth's history comprising the Triassic, Jurassic and Cretaceous Periods. The Mesozoic Era lasted from about 225 to 60 million years ago.

Metamorphosis The change, in an animal, from the larval form to the adult.

Migration The seasonal movement of certain animals from one place to another, in search of better food or breeding conditions.

Mushroom The spore-producing body of certain fungi.

Monotreme A primitive, egg-laying mammal found in Australasia. Two living species are known; the platypus and the spiny anteater or echidna.

Moss A lowly, spore-bearing green plant. Mosses occur on the bark of trees, on walls, in ditches and other damp places.

Nectar A sugary fluid produced by many flowering plants, which is attractive to insects.

Nocturnal Active at night.

Nut A hard, woody structure containing a plant seed. For example a hazel nut.

Nymph A stage in the life cycle of certain insects. Nymphs resemble the adults except that their wings are not usually developed.

Ovule The egg-containing structure of a seed-bearing plant.

Parasite A plant or animal which lives in, or on, another organism and from which it gets its food.

Petal One of the parts of a flower. Petals are usually brightly coloured to attract insects to visit the flower.

Plankton Microscopic animals and plants which float in the upper layers of the world's oceans and lakes.

Pollination The transfer of pollen from the anthers to the stigma.

Pupa The stage between the larva and the adult in certain insects such as butterflies and moths. Also known as a chrysalis.

Predator An animal which hunts other animals for food.

Reptile A cold-blooded, scaly vertebrate animal descended from amphibians. Most reptiles lay shelled eggs. Common reptiles include snakes, lizards, crocodiles and tortoises.

Rodent One of a group of gnawing mammals. Rodents are the most widespread of all mammals and include squirrels, rats, mice and beavers.

Scale One of the flat, horny plates covering the bodies of certain animals such as reptiles and fishes.

Spore Reproductive structure produced by certain lowly animals and by many groups of plants. Spores are released from the parent and eventually give rise to new individuals.

Stigma The region of the female flower parts on which pollen must land during fertilization.

Succulent A plant which stores water within its tissues, for instance a cactus.

Temperate zone The regions of the world lying between the tropical zones and the polar regions.

Territory The part of a habitat which an animal defends against others of its species, usually for the purpose of breeding.

Thorax One of the parts of the body of an insect. The thorax is composed of three segments and bears the legs and wings.

Toadstool The spore-producing body of certain fungi.

Tropical zone The regions of the world lying to the immediate north and south of the equator.

Venom The poisonous liquid produced by certain animals such as some species of snakes and spiders. Venom is injected into the victim by biting or stinging, sometimes as a means of protection and sometimes to render the prey harmless before being eaten.

Vertebrate One of the group of mainly large animals which possess a backbone. Vertebrates include fishes, amphibians, reptiles, birds and mammals.

Index

Numbers in italics refer to illustrations.